Editors: W. Foerst and H. Grünewald

Ch T

Chemical Topics for Students

3

J. Jander - Ch. Lafrenz

Ionizing Solvents

John Wiley & Sons Ltd.
Verlag Chemie

J. Jander − Ch. Lafrenz
Ionizing Solvents

Chemical Topics for Students

3 Edited by
Wilhelm Foerst
and Helmut Grünewald

J. Jander – Ch. Lafrenz

Ionizing Solvents

Translated by Express Translation Service, London

John Wiley & Sons Ltd 1970
Verlag Chemie

TITLE OF THE ORIGINAL GERMAN PUBLICATION:
WASSERÄHNLICHE LÖSUNGSMITTEL
EDITED BY WILHELM FOERST AND HELMUT GRÜNEWALD
© BY VERLAG CHEMIE, GMBH, WEINHEIM/BERGSTR.,1968

With 7 figures and 33 tables

LIBRARY OF CONGRESS CATALOG CARD NUMBER: 77-116653

ISBN 0 471 43970 3 John Wiley & Sons Ltd

Printed in Germany
©1970 Verlag Chemie GmbH. Weinheim/Bergstr.

Composition· Mitterweger KG, Heidelberg. Printer: Colordruck, Heidelberg
Creation of pocket: Hanswalter Herrbold, Opladen
Illustration: Gert Nemela, Schwetzingen

Editors' Preface

Modern chemistry can no longer be taught and studied simply as organic. inorganic, and physical chemistry. First, the boundaries between these areas have become fluid; and, secondly, neighbouring fields — in particular, some in medicine and biology — have been added but cannot be arranged in this classification.

It is difficult to allow for this variety of matter in a normal chemistry course, for it is hardly possible to devote special lectures at a University or Technical College to each of the important partial areas. On the other hand, it would be excessive to demand that even an advanced student should read extensive monographs in order to obtain his first introduction to a subject.

We have therefore created the series of "Chemical Topics" whose volumes are intended to be, as it were, primers that impart the most important facts from a single field. Those who wish to pursue a theme further can do so from the literature cited.

The "Chemical Topics" are intended to be read in conjunction with the basic textbooks. They extend the coverage of the latter, increase the penetration, and lead the reader to the latest position in a given field. It is hoped to treat all the important partial areas during the coming years, creating in the course of time a kind of encyclopaedia of modern chemistry.

W. Foerst
H. Grunewald

Preface

This book is intended mainly as a short textbook for advanced students. Based on lectures, it will try to provide an introduction to chemistry in water-like ionizing solvents, a field that is not treated in a coherent manner in the existing textbooks. However, it also has the aim of stimulating the further study of this field. It seemed to us that the best way of doing this would be to refer the reader to recent detailed reviews and books. Original papers are cited only where reviews are at present sketchy or non-existent. The reader who requires further details can find references to the original publications in the books and reviews. As in many short textbooks, because the literature is presented in this way, the name of an author may not appear in connection with his work. We hope that all our colleagues will forgive this omission in the interests of subject matter.

Berlin   J. Jander
January 1968   C. Lafrenz

Table of Contents

1. General Notes on Water and Water-Like Solvents

1.1. Characteristic Properties and Definition

Chemists often tend to carry out reactions in water without considering in full how the result is influenced by the properties of water. The importance of water as a solvent can be seen from the fact that other, much less widely used solvents are described as "nonaqueous" or "water-like". This is due, not only to the abundance of water, but also to the fact that it possesses a number of special properties:

1. Excellent solvent power for many inorganic and organic substances

2. A very low conductivity (very little self-ionization: $2 H_2O \rightleftharpoons H_3O^+ + OH^-$) in the pure state

3. Ionizing power (ionic reactions)

4. Ability to form adducts (hydrates, hydration)

5. Possibility of neutralization reactions of acids and bases with formation of water

6. Conversion of nonelectrolytes (ansolvo acids, ansolvo bases) into acids or bases by combination with the ions of water

7. Ability to hydrolyse

8. Possibility of amphoteric behaviour

This list also shows why organic solvents such as benzene, hydrocarbons, and chloroform may be regarded as "nonaqueous", but not as "water-like" solvents, since they possess few or none of these properties. On the other hand, liquid hydrogen fluoride, liquid ammonia, and anhydrous sulphuric acid, are "water-like" solvents. Because of the most important of their properties, these solvents are also known as "ionizing" solvents.

Of all the "nonaqueous" solvents, only the "water-like" or "ionizing" solvents will be dealt with in the present monograph, and a selection that the authors consider typical will be discussed.

1.2. Physico-Chemical Aspects

Acid-Base Concepts for Protonic Solvents

Let us consider the physico-chemical quantities and laws that give water and the ionizing solvents their special position and determine their usefulness.

The usefulness of a liquid solvent depends first of all on its melting point and its boiling point. The temperature range in which water is normally used extends from 0 to 100 °C. With liquid ammonia it is possible to work at temperatures between −77.7 and −33.35 °C, while anhydrous sulphuric acid allows reactions above 100 °C (sulphuric acid boils at 290 - 317 °C). The choice of a nonaqueous solvent is not, however, determined only by the reaction temperature. Other factors that must be considered include solvent power and acidity or basicity. In the case of liquid ammonia, for instance, the basicity of the solvent far outweighs the technically rather inconvenient temperature range.

Other important physical quantities are the heats of solution and of vaporization, since these provide information about association of the solvent, which is partly due to dipole forces and partly to hydrogen bonds. For example, the heats of vaporization (generally expressed by the Trouton constant $\Delta H/T$, where ΔH = heat of vaporization at the boiling point and T = absolute boiling point) of water and ionizing solvents are considerably greater than those of organic, nonionizing solvents, the Trouton constant of which has the normal value of 21.5. Dipole forces always occur when the charge distribution in the molecule is unsymmetrical (e.g. in iodine chloride the chlorine is the negative pole and the iodine the positive pole of the molecule). The polarity of the solvent molecules is largely responsible for the dissolution of polar substances. Thus water with its strongly dipolar character cannot dissolve hydrocarbons or other apolar compounds in appreciable quantities. Polar organic substances, on the other hand, such as sugars or lower alcohols are soluble in water.

The solubility of a substance in a solvent is strongly dependent on several energy terms: the lattice energy of the crystal to be dissolved (which opposes dissolution), the solvation energy (which increases solubility), and the forces of attraction between the solvent molecules (which oppose dissolution). The lattice energy is the most important of these, and is the reason e.g. for the fact that the solubility of ionic halides in liquid ammonia decreases from the iodides to the fluorides. The interaction energies between the ions and molecules of a solution are appreciably influenced by the dielectric constant of the solvent. According to Coulomb's law, the force of attraction K between two point electric charges is given by

$$K = \frac{e_1 \cdot e_2}{D \cdot d^2}$$

(where e = charge, d = distance between the charge carriers, D = dielectric constant).
Thus for high values of D the Coulomb attraction is small. Ionizing solvents generally
have high dielectric constants (e.g. water 80, liquid ammonia 25 at $-77.6\,^{\circ}C$, 22.38 at
$-33.35\,^{\circ}C$). A high value of D decreases the Coulomb attraction between the ions of
the substance being dissolved, and so reduces the energy required for the dissolution
process. In solution, the cations and anions of the solute are enveloped in the dipole
molecules of the solvent (hydration, solvation), corresponding to their charge. According to the Born equation

$$E_{solv} = \frac{n^2}{2r}(1 - D)$$

(where E_{solv} = solvation energy, n = charge on the ion, r = ionic radius), the solvation
energy becomes increasingly negative with increasing dielectric constant of the solvent. Thus unlike organic solvents, which have low dielectric constants, ionizing
solvents can dissolve ionic substances by dissociating them into ions, as can be seen
from the electrical conductivity of these solutions.
The mobility of an ion, which can be calculated from its equivalent conductivity at
infinite dilution, can differ widely in different solvents. This is due both to the different degrees of solvation of the ion and to differences in the viscosities of the
solvents. The mobility of an ion is much lower in viscous (e.g. concentrated sulphuric
acid) than in very fluid solvents (e.g. water or liquid ammonia) (Walden's law). High
viscosities can cause difficulties in precipitation operations, crystallizations, or fil-
trations, even though they do not influence the chemical act.
It is also important to know the redox potential of a solvent. The solvent chosen
must be oxidized or reduced by the substances reacting in it, or any such redox
reactions must at least be very slow. Thus reducing agents that reduce hydrogen ions
to hydrogen cannot be used in water, while liquid ammonia reacts with strong
oxidizing agents to form hydrazine or nitrogen.
The classical Arrhenius theory of acids and bases cannot be used as a basis for a definition of acids and bases in ionizing solvents in general, since it refers only to water.
Though the proton, as in water, is the source of acidic properties in protonic ionizing
solvents such as hydrogen fluoride, hydrogen chloride, sulphuric acid, and liquid
ammonia, the hydroxide ion as a source of basic properties is confined to water.
For this reason the Brønsted theory, in which both acids and bases are defined in

terms of the proton (acid = proton donor, base = proton acceptor), is often used for protonic ionizing solvents.

Another method of defining acids and bases in protonic ionizing solvents seems more useful to us, since, like the Arrhenius theory for water, it is based on the solvent. This is the solvent theory of acids and bases. It is a logical generalization of the Arrhenius theory, since it is derived from the self-ionization of the particular solvent in question.

Self-ionization of water: $$2\,H_2O \rightleftharpoons H_3O^+ + OH^-$$

of liquid ammonia: $$2\,NH_3 \rightleftharpoons NH_4^+ + NH_2^-$$

of liquid hydrogen fluoride: $$2\,HF \rightleftharpoons H_2F^+ + F^-$$

Just as according to Arrhenius all substances that increase the proton concentration (i.e. the concentration of the solvent cation) are acids and all substances that increase the hydroxide ion concentration (i.e. the concentration of the solvent anion) are bases, so in the solvent theory of acids and bases all substances that increase the concentration of the solvent cation are acids and all substances that increase the concentration of the solvent anion are bases. The acidic and basic actions defined in this manner can arise in two ways. An acidic action (increase in the concentration of the solvent cation) results from the release of solvent cations (solvo acids) or from the capture of solvent anions (ansolvo acids).

Solvo acid in water: $$\left[H_3O\right] Cl \rightleftharpoons H_3O^+ + Cl^-$$

Solvo acid in liquid ammonia: $$NH_4Cl \rightleftharpoons NH_4^+ + Cl^-$$

Ansolvo acid in liquid hydrogen fluoride: $$BF_3 + 2\,HF \rightleftharpoons H_2F^+ + BF_4^-$$

Similarly, a basic action (increase in the concentration of solvent anions) results from the release of solvent anions (solvo bases) or from the capture of solvent cations (ansolvo bases).

Solvo base in water: $$NaOH \rightleftharpoons Na^+ + OH^-$$

Solvo base in liquid ammonia: $$KNH_2 \rightleftharpoons K^+ + NH_2^-$$

Solvo base in liquid hydrogen fluoride: $$KF \rightleftharpoons K^+ + F^-$$

Ansolvo base in water: $$NH_3 + H_2O \rightleftharpoons NH_4^+ + OH^-$$

Ansolvo base in liquid hydrogen fluoride: $$CH_3CO_2H + HF$$
$$\rightleftharpoons HCH_3CO_2H^+ + F^-$$

It can be seen that solvo acids and bases must have ions in common with the solvent. Ansolvo acids and bases must act on the dissociation equilibrium of the solvent in such a way as to cause the formation of further solvent ions.

Thus in the solvent theory, the question of whether a substance is to be regarded as an acid, a base, or a salt in a given solvent depends on the solvent. The ammonium salts, which are well known as salts in aqueous solution, are solvo acids in liquid ammonia; the metal amides, on the other hand, release amide ions in liquid ammonia. and are therefore solvo bases in this solvent.

The solvent theory of acids and bases is so general that it can also be used for the definition of acids and bases in solvents containing no protons. This will be discussed later (Section 7.1). The solvent theory will first be used as a classifying principle for protonic solvents.

Strong acids[*] such as hydrochloric acid, nitric acid, or perchloric acid exhibit no difference in strength at the same concentration in water owing to complete dissociation. Thus water as a solvent has a levelling effect on the strengths of these acids. The same is true of strong bases. To differentiate the strengths of acids and Strong acids[*] such as hydrochloric acid, nitric acid, or perchloric acid exhibit no between the various acids are more pronounced in glacial acetic acid, which according to its acidity[**] is converted into the acetonium ion $CH_3C(OH)_2{}^+$ only by stronger acids. On the other hand, CH_3COOH is levelling solvent for bases, such as OH^-, CN^-, and CH_3O^-, since the base CH_3COO^- is formed preferentially in these cases. The situation is reversed in liquid ammonia, which according to its basicity[**] has a levelling effect on acids (perferential formation of the ammonium ion NH_4^+) but a differentiating effect on bases.

1.3. Possibility of Chemical Variation by the Use of Different Solvents

A few examples will now be given to show how some reactions can be varied by the choice of solvent. These reaction types will be dealt with in greater detail in the discussion of the individual solvents.

The formation of precipitates, for example, is a function of the solubility of the substances involved in the reaction. Owing to the differences in the solubilities of barium chloride and silver chloride in water and in liquid ammonia, the reaction

$$2\,AgNO_3 \;+\; BaCl_2 \xrightleftharpoons[H_2O]{NH_3} 2\,AgCl \;+\; Ba\,(NO_3)_2$$

---

[*] We use the terms ,,acid strength, strong acid, weak acid" to caracterize the value of dissociation constant

[**] We use the terms ,,acidity, basicity" to caracterize the value of protonactivity

leads in water to the formation of silver chloride, and in liquid ammonia to the precipitation of barium chloride. Calcium chloride, sodium hydroxide, and ammonium sulphate, which are readily soluble in water, are insoluble in liquid ammonia, whereas copper (I) iodide is readily soluble in liquid ammonia, but insoluble in water. The water-soluble copper (II) sulphate, on the other hand, can be precipitated from glacial acetic acid.

Another reaction well known in the aqueous system is the formation of salts by neutralization of acids with bases. This type of reaction depends on the use of a solvent in which both the acid and the base can exist. For example, the sodium salt of urea $H_2NCONHNa$ cannot be prepared in water, since even the strongest base cannot remove a proton from urea in water. This salt can however be prepared in liquid ammonia. in which the more protophilic amide ion is available as a base. Sodium amide and urea react in liquid ammonia to form $NH_2CONHNa$. The well known nitronium perchlorate NO_2ClO_4 also cannot be prepared in water because of hydrolysis:

$$NO_2^+ + 2\,H_2O \longrightarrow NO_2OH + H_3O^+.$$

However, NO_2OH (=HNO_3) can be converted into NO_2^+ with concentrated sulphuric acid.

The decomposition of compounds by solvent molecules into acids and bases is known by the general name of solvolysis. Depending on the solvent, one speaks of hydrolysis (water), ammonolysis (liquid ammonia), acetolysis (glacial acetic acid), etc. The solvolysis of the same compound in different solvents yields different products. For example, solvolysis of sulphuryl chloride in water leads to sulphuric acid and hydrochloric acid, while solvolysis in liquid ammonia proceeds in accordance with the equation

$$SO_2Cl_2 + 4\,NH_3 \longrightarrow SO_2(NH_2)_2 + 2\,NH_4^+ + 2\,Cl^-.$$

Solvolysis in glacial acetic acid leads to $SO_2(OOCCH_3)_2$, $CH_3C(OH)_2^+$, and chloride ions.

Finally, an important process in the dissolution of compounds is solvation. This process may proceed differently in different solvents, and often leads to definite compounds known as solvates. These include the numerous hydrates, such as $[Cu(H_2O)_4]$ SiF_6, $[Cu(H_2O)_4]SO_4 \cdot H_2O$, and the alums. Many corresponding ammonia compounds (ammoniates, ammines) are known, though there are also some differences. For example, barium reacts with water to form H_2

$$Ba + 2\,H_2O \longrightarrow Ba^{2+} + 2\,OH^- + H_2,$$

whereas addition of the metal to liquid ammonia yields a blue solution, from which the bronze-coloured $Ba(NH_3)_6$ can be obtained by careful evaporation of the solvent. Solvation, in which the solvent molecules are bound by ion-dipole forces, hydrogen bonds, or coordinate bonds, frequently precedes solvolysis.

1.4. Bibliography

H.H. Sisler: Chemistry in Non-Aqueous Solvents. Reinhold, New York 1961
L.F. Audrieth u. J. Kleinberg: Non-Aqueous Solvents, Wiley, New York 1953, Chapter 1
J.Jander: Anorganische und allgemeine Chemie in flüssigem Ammoniak, in G. Jander, H. Spandau u. C.C. Addison: Chemie in nichtwässrigen ionisierenden Lösungsmitteln. Vieweg, Braunschweig 1966,Vol. I,1

2. Liquid Ammonia

The use and investigation of liquid ammonia as a solvent began more than 60 years ago with the work of Franklin, Kraus, and Cady. The 4000 or so publications that have appeared thus far make liquid ammonia the most thoroughly investigated of the ionizing solvents.

2.1. Physico-Chemical Properties of Ammonia

Ammonia in equilibrium with its saturated vapour melts at -77.7 °C (ND_3 melts at -74.1 °C), and boils at -33.4 °C (the boiling point of ND_3 is -31.05 °C). The critical pressure is reported to be 112.3 atm, the critical temperature 132.4 °C, and the critical density 0.2362 g/ml.

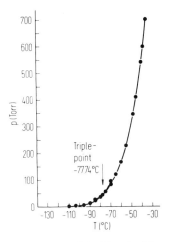

Fig. 1. Vapour pressure of liquid and solid ammonia

The vapour pressure curve is shown in Figure 1. Between -73.84 and -31.51 °C, the vapour pressure can be calculated from the equation

$$\log p = -1612.500/T - 0.012311\ T + 0.000012521\ T^2 + 11.83997 \quad (47)$$

$(T = °K).$

The average value found for the heat of fusion from experimental and calculated values is $\Delta H_f = 1376$ cal/mole.

The average heat of vaporization at the boiling point and at atmospheric pressure is $\Delta H_v = 5598$ cal/mole.

Density at the boiling point : 0.6814 g/cm^3.
Viscosity at the boiling point : 2647 x 10^{-6} poise.

The dielectric constant D of liquid ammonia is considerably lower than that of water (Table 1).

Table 1. Dielectric constant D of liquid ammonia at various temperatures.

Temp. (°C)	D	Temp. (°C)	D
−166.1	3.36	−33.35	22.38
− 79.0	4.83	+ 5	18.94 ± 0.05
− 77.6	25	+15	17.82 ± 0.01
− 50	24.3		

The dipole moment of liquid ammonia at −50 °C is 1.69 x 10^{-18} dyne$^{\frac{1}{2}}$cm^2. Liquid ammonia is dissociated to a small extent:

$$2\,NH_3 \rightleftharpoons NH_4^+ + NH_2^-.$$

The ionic product of about 10^{-29} is even smaller than that of alcohol (approx. 10^{-20} Table 2 shows how the pH range and the neutral point of protonic ionizing solvents vary with the ionic product.

Table 2. Ionic products, pH ranges, and neutral points of some solvents

Solvent	Ionic product	pH-range	Neutral point
CH_3COOH	10^{-13}	0–13	6.5
H_2O	10^{-14}	0–14	7
C_2H_5OH	10^{-20}	0–20	10
NH_3	10^{-29}	0–29	14.5

It is very difficult to measure the conductivity of liquid ammonia since even very small quantities of ionic impurities lead to relatively large errors. The specific conductivity at −34 °C is 1 x 10^{-11} Ω^{-1}cm^{-1}.

Comparison of the melting and boiling points, the heats of fusion and vaporization, the Trouton constant, the viscosity, and other physico-chemical data shows that liquid ammonia, like water, alcohol, and glacial acetic acid, is associated. Hydrogen bonds having a calculated energy of 5.1 kcal/mole exist between the ammonia molecules.

2.2. Physico-Chemical Properties of Solutions of Inorganic and Organic Substances in Liquid Ammonia

2.2.1. Solubilities

Table 3 shows a selection of inorganic substances whose solubilities in liquid ammonia have been determined.

Table 3. Solubilities of inorganic substances in liquid ammonia (i = insoluble, v.sl. = very slightly soluble, sl. = slight soluble, s = soluble, f.a. = fairly abundantly soluble, a = abundantly soluble)

Substance	Solubility (g/100 g NH_3)	Temp. °C	Substance	Solubility (g/100 g NH_3)	Temp. °C
AgBr	5.29	+25	K_2CO_3	i.	+25
AgCl	0.83	+25	KCl	0.04	+25
AgF	sl.	—	$KClO_3$	2.52	+25
AgI	206.84	+25	$KClO_4$	s	—
$AgNO_3$	86.04	+25	KF	sl.	—
$AlCl_3$	sl.	−30	KI	a.	—
BF_3	about 14.7	−33	KNH_2	3.6	+25
$Ba(NO_3)_2$	s	−50	KNO_2	a.	—
$BaCl_2$	i.	+25	KNO_3	10.4	+25
$CaBr_2$	v.sl.	−33	KOH	i.	—
$CaCl_2$	0.08	+19	KSCN	a.	+20
$Ca(ClO_4)_2$	52.6	−33	K_2SO_4	i.	—
CaI_2	3.85	0	LiBr	0.169	+20
$Ca(NH_2)_2$	v.sl.	—	LiCl	1.43	0
$Ca(NO_3)_2$	80.22	+25	LiI	a.	—
$CaSO_4$	i.	—	$LiNH_2$	v.sl.	—
CdF_2	i.	—	$LiNO_3$	243.66	+25
CsBr	4.38	0	$MgBr_2$	v.sl.	−33

Table 3. Continuation

Substance	Solubility (g/100 g NH$_3$)	Temp. °C	Substance	Solubility (g/100 g NH$_3$)	Temp. °C
CsCl	0.381	0	MgCl$_2$	i.	+25
CsI	60.28	–	NaBr	137.95	+25
CsNH$_2$	f.a.	–	NaCN	40	+10
CuI	about 34	+25	Na$_2$CO$_3$	i.	+25
CuSO$_4$	i.	–	NaCl	4.2 ± 0.2	+25
Fe(CN)$_2$	s	–	NaClO$_3$	f.a.	–
H$_2$O	a.	–	NaF	0.35	+25
Hg(CN)$_2$	a.	–	NaI	161.9	+25
HgI$_2$	(Reaction)	+25	NaNH$_2$	i.	+25
Hg$_2$I$_2$	i.	–	NaN$_3$	f.a.	–
KBr	13.5	+25	NaNO$_2$	a.	<–33,5
KCN	5.5	–33	NaNO$_3$	97.6	+25
NaOH	i.	+20	NH$_4$HCO$_3$	i.	+25
NaSCN	205.5	+25	NH$_4$F	i.	–
Na$_2$SO$_4$	i.	+25	NH$_4$I	368.5	+25
NiSO$_4$	i.	+20	NH$_4$N$_3$	about 42	–33
Ni(NH$_2$)$_2$	i.	+25	NH$_4$NO$_3$	390	+25
NH$_4$Br	237.9	+25	(NH$_4$)$_3$PO$_4$	i.	–
(NH$_4$)$_2$CO$_3$	i.	–	NH$_4$SCN	312	+25
NH$_4$Cl	102.5	+25	(NH$_4$)$_2$SO$_4$	i.	–30 to
NH$_4$ClO$_4$	137.93	+25	N$_2$H$_4$	f.a.	–10 to

It can be deduced from this list and other solubility studies, as well as from the remar in Section 1.2 concerning solubility, that the halides (Cl$^-$ < Br$^-$ < I$^-$), cyanides, thiocyanates, nitrates, and nitrites are generally readily soluble in liquid ammonia whereas fluorides, hydroxides, oxides, sulphides, sulphites, sulphates, phosphates, carbonates, and chromates exhibit little or no solubility. For the alkali metal salts of an anion, the solubility decreases in the order Li$^+$ > Na$^+$ > K$^+$ > Rb$^+$ > Cs$^+$. The ammonium salts, i.e. the solvo acids in liquid ammonia, are mostly more readily soluble than the corresponding alkali metal salts owing to the possibility of hydrogen bonding. The solubility of the silver halides (due to complex formation) is particularly interesting.

Liquid ammonia is also widely used as a solvent for organic compounds.

Amines: Simple amines are readily soluble. Primary amines are more soluble than secondary, and these in turn are more so than tertiary amines.

Amides and amidines: the simple members of this class of compounds are readily soluble.

Nitrogen heterocycles: Pyridine, quinoline, indole, pyrrole, carbazole, and simple triazoles and tetrazoles are readily soluble.

Alcohols: Simple and polyfunctional alcohols are miscible with liquid ammonia. Phenols are also soluble.

Carboxylic acids: The ammonium salts of the lower carboxylic acids are soluble.

Esters: The simple esters are readily soluble. The solubility decreases with the size of the alkyl groups.

Aldehydes and ketones: These are moderately soluble. Reaction occurs with ammonia.

Ethers: Diethyl ether is moderately soluble. Ethers having higher molecular weights are not very soluble.

Sulphonic acids: The ammonium salts formed are soluble.

Hydrocarbons: Alkanes are insoluble, while alkenes and alkynes are slightly soluble. Benzene dissolves readily.

In all these classes of substances, the solubility decrease with increasing size of the hydrocarbon residue. In general, the solvent power of liquid ammonia for organic compounds is better than that of water.

2.2.2. Vapour Pressures of Solutions

Many measurements have been carried out on the changes in vapour pressure that accompany the dissolution of substances in liquid ammonia. Solutions of readily soluble substances such as ammonium thiocyanate, ammonium nitrate, and lithium nitrate naturally have very low vapour pressures; for example, the vapour pressure of a saturated solution of ammonium thiocyanate is 1 mm Hg at -78 °C, 21 mm Hg at -34 °C and 316 mm Hg at $+30$ °C. The vapour pressure of saturated ammonium nitrate solution at 0 °C is 364 mmHg, while solutions of 43.24 mole-% of salt at 20 °C give a value of 760 mmHg. These solutions, which were first investigated by Divers in 1873/74, are now known as Divers solutions. Solutions containing ammonium thiocyanate and ammonium nitrate in a ratio of 3:1 have even lower vapour pressures, a value of only 177 mmHg being obtained for a saturated solution at +20 °C. A saturated solution of $LiNO_3$ has a vapour pressure of 155 mmHg at +20 °C. Ammoniu thiocyanate — ammonium nitrate mixtures or lithium nitrate can therefore be used for the absorption of ammonia from gas mixtures. Absorption vessels made of iron, machining steel or chrome-nickel alloys are not attacked by solutions of lithium nitrate in liquid ammonia.

2.2.3. Ammonia Adducts

Though the many known ammonia adducts (now known as ammines, and formerly as ammoniates) cannot be discussed in detail, brief mention must be made of this adduct formation, since it precedes the dissolution processes and may be a precursor of solvolysis, which will be discussed later.

Ammonia has a much greater ability than water to add on to compounds that are already chemically saturated. It is possible to distinguish various types of adducts, though no sharp separating line can be drawn between them. If compounds having a pronounced salt character, i.e. having an ionic lattice, take up ammonia, the products are referred to as interstitial compounds, since the ammonia is incorporated into the lattice. With the partial negative charge on the nitrogen (the ammonia molecule is a dipole) it mainly attacks the cations and separates them from the surrounding anions. This process is a precursor of dissolution. An interesting variant of the inclusion of ammonia is found in the polyanionic salts, e.g. Na_4Pb_9 or Na_4Sn_9, which may be regarded as alloys in the absence of ammonia, whereas they form anionic lattices on incorporation of ammonia. The adduct formation must be accompanied in every case by expansion of the lattice, and the energy involved ("work of expansion") must be less than the energy liberated on incorporation of the ammonia molecules. Owing to the high work of expansion required, many fluorides are unable to form adducts directly with ammonia; in such cases, ammonia adducts can only be formed via the hydrates.

Compounds having molecular or layer lattices, in which the central atom of the molecule dissolved out of the lattice is coordinately unsaturated, add on one or two molecules of ammonia with retention of the bonds between the metal and the nonmetal to give addition compounds, which generally have a molecular lattice. This type of reaction is found with e.g. boron trifluoride and tin tetrachloride. A layer lattice such as that of aluminium chloride is converted into a typical molecular lattice, in this case with formation of $Cl_3Al(NH_3)$ molecules. The central atoms of the resulting adduct molecules frequently exhibit higher coordination numbers in the lattice; for example $Cl_2Cd(NH_3)_2$ consists of chains of polyhedra, while Cl_2Hg $(NH_3)_2$ has a three-dimensional network structure.

The ammonia adducts of the transition metals occupy a special place among the interstitial and addition compounds because of their stability. In addition to the incorporation of ammonia as a result of dipole forces, strong electronic interactions must also be assumed to occur in this case. The relatively high polarizability of the ammonia molecule favours complex formation. Thus water ligands, because of their low polarizability, are readily displaced by ammonia, though water has the higher dipole moment. Many complexes of this type are stable even in aqueous solution. Examples of such complexes are $Hg(NH_3)_2{}^{2+}$, $Pt(NH_3)_4{}^{2+}$, $Cu(NH_3)_4{}^{2+}$, $Cr(NH_3)_6{}^{3+}$, $Co(NH_3)_6{}^{3+}$, and $Ni(NH_3)_6{}^{2+}$.

The composition of the adducts can be recognized not only by analysis of the more stable compounds, but also by vapour pressure measurements. In these measurements, ammonia is gradually removed from preparations having high ammonia contents, and the resulting vapour pressures are measured (tensimetric degradation). The vapour pressure curves obtained contain steps corresponding to the characteristic pressures of the adducts. Another method is the plotting of melting diagrams.

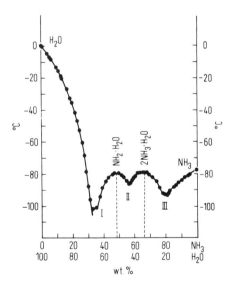

Fig. 2. Melting diagram of the ammonia-water system

For example, Figure 2 shows the melting diagram of the water-ammonia system, from which the existence of two ammines of water can be recognized.

2.2.4. Electrochemical and Osmotic Studies

Various methods can be used to determine the form in which a substance exists in a solvent; these include conductivity measurements, molecular weight determinations, and potentiometric measurements. Electrical conductivity directly demonstrates the presence of ions, while the degree of dissociation can be deduced from the concentration-dependence of the equivalent conductivity. The curves obtained for solutions in liquid ammonia are qualitatively similar to those obtained in aqueous solutions. The equivalent conductivity increases with increasing dilution (Ostwald's dilution law); even substances that are strong electrolytes in aqueous chemistry, such as potassium iodide and copper(II) nitrate, exhibit increasing equivalent conductivity with increasing dilution, and are therefore not completely dissociated. On the other hand, quantitative deviations from Ostwald's dilution law are observed; however, these cannot be discussed here. The incomplete dissociation of strong electrolytes in liquid ammonia is due to the lower dielectric constant of the solvent. Despite incomplete dissociation, however, the equivalent conductivity in liquid ammonia at $-40\ ^\circ$C is greater than that in water at $+80\ ^\circ$C (Table 4), i.e. the migration velocity of the individual ion is greater in liquid ammonia than in water. This is partly due to the lower viscosity of ammonia at comparable temperatures (ammonia $-40\ ^\circ$C, water $+80\ ^\circ$C).

Table 4. Equivalent conductivities of the alkali metal ions $(\Omega^{-1}\ cm^2\ eqt^{-1})$

	$NH_3(-40\ ^\circ C)$	$H_2O\ (+80\ ^\circ C)$
Li^+	121	80
Na^+	131	110
K^+	170	160

The equivalent conductivities (ionic mobilities) of the ammonium and amide ions lie in the same range as the mobilities of other ions. Thus the proton is not such a special case in liquid ammonia as in water. The special conduction mechanism that occurs in water is prevented by the stronger bonding of the proton in the ammonia molecule. Conductivity measurements can be used to determine solubilities and activity coefficients or to investigate the kinetics of chemical reactions. The equivalent conductivity of potassium ozonide in liquid ammonia ($34\ \Omega^{-1}\ cm^2\ eqt^{-1}$ at $-50\ ^\circ$C) shows that this substance has an ionic structure $K^+O_3^-$. Boron trifluoride-ammonia was found by conductivity measurements in liquid ammonia to be a weak acid $BF_3.NH_2^-NH_4^+$.

with a degree of dissociation of 1%. Conductivity measurements in the form of conductimetric titrations can be used for the quantitative investigation of a reaction course, for example in the reaction of tin tetrahydride with sodium, which leads to the formation of sodium trihydridostannate, $Na[SnH_3]$.

The lower degree of dissociation of salts in liquid ammonia than in water can be confirmed by molecular weight determinations. The calculated cryoscopic constant for liquid ammonia is 0.94 °C/mole, whereas the values found in practice are 0.94 to 1.7 °C/mole. Because of this anomalous cryoscopic behaviour, absolute molecular weight determinations by the freezing point method in liquid ammonia are of little value. The melting point determination can really be used only to check the change in the number of dissolved particles with concentration. Thus the cryoscopic constant for solutions of ammonium nitrate or potassium iodide decreases rapidly with increasing concentration (0.05 to 0.4 M). (prevalence of association over dissociation), and reaches a minimum in 0.4 M solution (maximum association). When the concentration is further increased, the value of the constant increases once more (increasing solvation, for explanation see later).

$$(KI)_m(NH_3)_n \underset{+nNH_3}{\overset{-nNH_3}{\rightleftharpoons}} (KI)_m \rightleftharpoons mKI \rightleftharpoons mK^+ + mI^-$$

Solvation   Association   Dissociation

The most reliable molecular weight determinations in liquid ammonia are obtained by measurement of the vapour pressure lowering, since no anomalies occur in this case. Determinations of this type have shown e.g. that diborane B_2H_6 is neither associated nor dissociated in liquid ammonia at −35 °C. An apparent decrease in the molecular weight with increasing concentration suggests that the compound $B_2H_6 \cdot 2NH_3$ that is actually present binds further ammonia molecules (increasing solvation) and so decreases the quantity of free solvent. The concentration of B_2H_6 consequently increases, and the molecular weight appears to decrease.

Further information about the state of substances in liquid ammonia was provided by investigations that led to the construction of an electromotive series for this solvent (Pleskov 1947). As in water, the redox potentials of the elements in liquid ammonia can be listed in electromotive series (Table 5), the normal potential being based on the normal hydrogen electrode in liquid ammonia; however, the normal hydrogen electrode is more difficult to produce experimentally in liquid ammonia than in water.

Table 5. Normal potentials in liquid ammonia and in water (V)

Metal	Against the hydrogen electrode		Against the rubidium electrode	
	in Liq. NH_3	in H_2O	in Liq. NH_3	in H_2O
Li/Li^+	−2.28	−3.01	−0.35	−0.03
K/K^+	−1.98	−2.92	−0.05	+0.06
Cs/Cs^+	−1.95	−2.92	−0.02	+0.06
Rb/Rb^+	−1.93	−2.98	0	0
Na/Na^+	−1.85	−2.713	+0.08	+0.27
Ca/Ca^{2+}	−1.64	−2.84	+0.29	+0.14
Zn/Zn^{2+}	−0.53	−0.763	+1.40	+2.22
Cd/Cd^{2+}	−0.20	−0.402	+1.73	+2.58
H_2/H^+	0	0	+1.93	+2.98
Pb/Pb^{2+}	+0.32	−0.126	+2.25	+2.85
Cu/Cu^+	+0.41	+0.52	+2.34	+3.50
Cu/Cu^{2+}	+0.43	+0.34	+2.36	+3.32
Hg/Hg^{2+}	+0.75	+0.85	+2.68	+3.84
Ag/Ag^+	+0.82	+0.799	+2.76	+3.78

It can be seen from the left hand part of Table 5, e.g. that lead is less noble than hydrogen in water, but more noble in liquid ammonia, or conversely hydrogen is more noble in water and less noble in liquid ammonia than lead. This difference in the character of hydrogen in different solvents is due to the difference in the solvation energy of the proton, which is higher in liquid ammonia than in water. On the other hand, because of their size and their low polarizability, rubidium and caesium ions have practically the same solvation energies in both solvents. Consequently, the difference in the normal potentials of rubidium (or caesium) in liquid ammonia and in water against the normal hydrogen elctrode must be equal to the energy of transfer of a proton from ammonia to water. This provides a direct relationship between the potentials against the normal hydrogen electrodes in ammonia and in water. Thus if the potentials are based, not on the normal hydrogen electrodes, but on the rubidium electrodes as shown in the right hand part of Table 5, the difference in the normal potential of a metal in ammonia and in water corresponds to the energy required for the transfer of the metal ion from water to ammonia, i.e. to the difference in the solvation energies. This then shows directly that the proton has a higher solvation energy in liquid ammonia, i.e. that it is less noble in this solvent.

An electromotive series (based on the normal hydrogen electrode in ammonia) can also be given for halide ions in liquid ammonia: I^- : + 1.3; Br^- : + 1.7; Cl^- : + 1.9 V. These values refer to ammono-acidic solutions.

It should be mentioned that the phenomenon of overvoltage has also been observed in liquid ammonia. For example, an overvoltage of 1.1 to 1.2 V has been found for hydrogen (electrolysis of ammonium chloride) at lead electrodes. As is shown by the simultaneous anodic liberation of nitrogen and iodine during the electrolysis of ammonium iodide at platinum electrodes, nitrogen has an overvoltage of approximately 1.2 V on platinum. The redox potential NH_3/N_2 in ammono-acid solutions at +25 °C is reported to +0.04 V, but as is found by experiment, approaches the potential of iodine.

The hydrogen and nitrogen overvoltages are very important for redox reactions in liquid ammonia (Section 2.6). If the pairs $H_2/NH_4{}^+$ and NH_3/N_2 were completely reversible, i.e. without overvoltages, the only redox systems that would be soluble in liquid ammonia without decomposition of the solvent would be those whose normal potentials in ammono-acid solution are between 0 and +0.04 V. Because of these overvoltages, however, the usable redox range extends from +1.0 to -1.0 V. In ammono-basic solutions, only redox systems whose normal potentials are between −1.59 V (H_2) and −1.55 V (N_2) would be soluble without decomposition in the absence of overvoltage. However, the overvoltage in ammono-basic solutions makes it possible to use the range between −0.6 and −2.6 V. During electrolyses, therefore, NH_4^+ ions are discharged at the cathode only at relatively low voltages. This permits e.g. the reduction of base metals such as beryllium or the alkali metals (to form the well known blue solutions). In the absence of the nitrogen overvoltage, the liberation of nitrogen would be the most strongly favoured process at the anode. Though the overvoltage permits the liberation of iodine, the liberation of the other halogens (which can occur directly during electrolysis in water) and of oxygen is impossible. Stronger anodic oxidations cannot therefore proceed in liquid ammonia.

Finally, it should be mentioned that potential measurements (potentiometric titrations) have been used for end-point indication in redox reactions in liquid ammonia, particularly for the investigation of the stoichiometry of reductions with alkali and alkaline earth metals dissolved in liquid ammonia (Section 2.7).

2.3. Acids and Bases

2.3.1. Ammono Acids

The self-ionization of liquid ammonia yields ammonium and amide ions.

$$2\,NH_3 \rightleftharpoons NH_4{}^+ + NH_2{}^-.$$

According to the solvent theory of acids and bases discussed in Chapter 1, therefore, the ammonium salts are regarded as acids and amides as bases in liquid ammonia, or more precisely, ammonium salts are solvo acids and amides are solvo bases in liquid ammonia. Ansolvo acids and ansolvo bases are not so common in this solvent (see below).

Examples of strong solvo acids in liquid ammonia are ammonium chloride, iodide, cyanide, thiocyanate, and tetrafluoroborate. Ammonium nitrite, nitrate, and azide are strong acids with oxidizing properties. The acidic character of ammonium salts can be seen in every case from the dissolution of base metals with liberation of hydrogen:

$$Mg + 2\,NH_4^+ \longrightarrow Mg^{2+} + 2\,NH_3 + H_2$$

$$Al + 3\,NH_4^+ \longrightarrow Al^{3+} + 3\,NH_3 + 3/2\,H_2.$$

Even sparingly soluble solvo acids such as ammonium fluoride and ammonium sulphate slowly corrode metals. The reactions of the acids with alloys are useful for the preparation of hydrides. For example, the reactions of silicon-magnesium and germanium-magnesium alloys in liquid ammonia yield silicon hydride and germanium hydride respectively:

$$Mg_2Si + 4\,NH_4^+ \longrightarrow 2\,Mg^{2+} + SiH_4 + 4\,NH_3$$

$$Mg_2Ge + 4\,NH_4^+ \longrightarrow 2\,Mg^{2+} + GeH_4 + 4\,NH_3.$$

As in the corresponding reactions in water, small quantities of higher silanes or germanes are formed as by-products. The reaction in liquid ammonia gives significantly better yields. Ammonium nitrate in liquid ammonia, like nitric acid in water, acts as an oxidizing acid and dissolves metals (M) in accordance with the equation:

$$2\,NH_4^+ + NO_3^- + 2\,M \longrightarrow 2\,M^+ + NO_2^- + H_2O + 2\,NH_3.$$

The dissolution of metal amides and imides by acids in liquid ammonia, corresponding to the dissolution of metal hydroxides or oxides by aqueous acids, is also observed; for example

$$Zn(NH_2)_2 + 2\,NH_4^+ \longrightarrow Zn^{2+} + 4\,NH_3$$

$$PbNH + 2\,NH_4^+ + 2\,I^- \longrightarrow PbI_2 + 3\,NH_3.$$

Iron(III) oxide, cobalt(III) oxide, antimony(V) oxide, tungsten(VI) oxide, and uranium (VI) oxide, on the other hand, are insoluble even in solutions of ammonium nitrate or ammonium chloride in liquid ammonia.

In addition to these strong acids many compounds are known to form salts with strong bases such as alkali metal amides or to react with alkali metals with salt formation and liberation of hydrogen, and must therefore be regarded as weak acids. Examples of such compounds are tin(IV) hydride ($=NH_4^+SnH_3^-$) and phosphine ($=NH_4^+PH_2^-$), which react with alkali metals to give salts having the composition $NaSnH_3$ and $NaPH_2$ respectively.

The amides and imides of the oxygen acids also have acidic properties in liquid ammonia, and are classed as solvo acids. Thus guanidine $NH_2C(=NH)NH_2$ and urea NH_2CONH_2, which are proton acceptors in water, act as proton donors in liquid ammonia and accordingly form salts with strong bases. Sulphamic acid, which is a strong monobasic acid in water, yields two protons in liquid ammonia due to its basicity with formation of the anion $^-OSO_2NH^-$. Sulphamide $H_2NSO_2NH_2$ behaves similarly. Acid amides (e.g. H_2NCN) form two series of salts (e.g. KHNCN and K_2NCN), while acid imides, as monobasic acids, can form only one series of salts.

Carbon dioxide and carbon disulphide act as ansolvo acids, i.e. they increase the concentration of solvent cations by combination with the solvent anions to form ammonium carbamate and thiocarbamate respectively:

$$CS_2 + 2\,NH_3 \longrightarrow NH_4^+ + [SC(=S)NH_2]^-$$

$$CO_2 + 2\,NH_3 \longrightarrow NH_4^+ + [OC(=O)NH_2]^-$$

Finally, it should again (cf. Chapter 1) be mentioned that owing to its strong basicity, liquid ammonia has a levelling effect on the strengths of acids, since it strengthens in particular the acids that are weak in water. This effect is particularly noticeable with the acids ammonium cyanide ($K=2 \times 10^{-3}$) and ammonium hydrogen sulphide ($K \approx 10^{-3}$), and is even more pronounced with compounds that are weak bases in water, such as urea and guanidine.

2.3.2. Ammono Bases

Whereas there are many medium and strong acids in liquid ammonia, there are only a few strong bases. Owing to the poor solubilities of lithium and sodium amides, potassium amide is most frequently used as a strong and readily soluble solvo base. The preparation of the alkali metal amides and of the alkaline earth metal amides, which are even stronger bases, is described in Section 2.7. The solutions of the bases are all weak electrolytes. Ansolvo bases are of no importance in liquid ammonia.

The precipitation reactions of the heavy metal ions by hydroxide ions with formation of hydroxides or oxides in water have a parallel in the precipitations from liquid ammonia as heavy metal amides, imides, or nitrides. Thus potassium amide precipitat silver amide from a solution of silver nitrate in liquid ammonia. Lead iodide similarly gives lead amide $Pb(NH_2)_2$, while mercury(II) iodide, bismuth iodide, and thallium (I) nitrate are converted into the corresponding nitrides. Many of these amides redissolve on further addition of potassium amide, i.e. they exhibit amphoteric character (see Section 2.3.5).

Imides and nitrides can be obtained from amides by heating under vacuum. The relationship is the same as that between hydroxides and oxides in the aqueous system, except that imides, which formally correspond to oxides, can lose further ammonia. There are also compounds that belong to both the aqueous and the ammonia system, such as Millon's base Hg_2NOH and the dark red, explosive bismuth oxide amide $BiONH_2$, which is formed from bismuth oxide iodide and potassium amide in liquid ammonia.

2.3.3. Neutralizations

Neutralization in liquid ammonia is also understandable from the analogy between acids and bases in the liquid ammonia and in the aqueous systems. Just as hydronium ions and hydroxide ions in water combine to form undissociated water, so undissociated ammonia molecules are formed from ammonium and amide ions. Typical neutralizations in liquid ammonia are the reaction of ammonium bromide with potassium amide:

$$NH_4^+ + Br^- + K^+ + NH_2^- \longrightarrow K^+ + Br^- + 2\,NH_3.$$

and the reaction of ammonium nitrate with potassium amide:

$$NH_4^+ + NO_3^- + K^+ + NH_2^- \longrightarrow K^+ + NO_3^- + 2\,NH_3.$$

The dissolution of zinc amide and lead imide by ammonium salts was mentioned in Section 2.3.1. Nitrides also dissolve in a similar manner:

$$Hg_3N_2 + 6\,NH_4^+ + 6\,NO_3^- \longrightarrow 3\,Hg^{2+} + 6\,NO_3^- + 8\,NH_3.$$

Nickel(II) amide, thallium and bismuth nitrides, and many other amides, imides and nitrides react similarly.

2.3.4. Acid-Base Indicators

In addition to the electrochemical end-point indications mentioned in Section 2.2, colour indicators can also be used for acid-base titrations in liquid ammonia. These indicators respond to changes in the NH_4^+ ion concentration, just as indicators in aqueous solution respond to changes in the H_3O^+ ion concentration. The only difference between the two systems is that owing to the lower ionic product and stronger basicity of liquid ammonia, the pH range in which an indicator changes colour is displaced, i.e. the acidic character of the indicator is strengthened. The dissociation scheme for nitrophenols

$$Ar(NO_2)_xOH \rightleftharpoons Ar(NO_2)_xO^- + H^+$$
colourless   yellow-orange

shows for example that the equilibrium is displaced to the left in proton-active solvents and to the right in not proton-active solvents. Thus the indicator is colourless in glacial acetic acid, bright yellow in water, and orange in liquid ammonia. Liquid ammonia exhibits such a basicity that the colour connot be discharged by the addition of ammonium ions, i.e. an acid. Many other indicators also show no appreciable colour change on acidification. Ammono-basic solutions, on the other hand, cause colour changes in many cases. For example, nitroanilines in liquid ammonia change from yellow to red or green on addition of potassium amide, thymol blue changes from yellow to blue, and phenolphthalein changes from colourless to red. An exception to the rule that colour indicators used in liquid ammonia can also be used in water is triphenylmethane, which is a sensitive indicator only in liquid ammonia and changes from colourless to deep red on addition of potassium amide. This red coloration is brought about even by the very sparingly soluble potassium hydroxide.

2.3.5. Amphoterism

It was mentioned in Section 2.3.2 that heavy metal amides precipitated from liquid ammonia by potassium amide in some cases redissolve on further addition of potassium amide. This process corresponds to the formation of hydroxo complexes, e.g. in the cases of zinc and aluminium, in the aqueous system. Amphoterism is thus also observed in liquid ammonia:

$$Al(OH)_3 + OH^- \longrightarrow [Al(OH)_4]^-$$
$$Al(OH)_3 + 3\,H_3O^+ \longrightarrow Al^{3+} + 6\,H_2O$$
$$Al(NH_2)_3 + NH_2^- \longrightarrow [Al(NH_2)_4]^-$$
$$Al(NH_2)_3 + 3\,NH_4^+ \longrightarrow Al^{3+} + 6\,NH_3.$$

Other amides also form soluble complexes; for example $AgNH_2$ gives $K[Ag(NH_2)_2]$, $Be(NH_2)_2$ gives $K[Be(NH_2)_3]$, $Zn(NH_2)_2$ gives $K_2[Zn(NH_2)_4]$, $Pb(NH_2)_2$ gives $K[Pb(NH_2)_4]$, and $Mg(NH_2)_2$ gives $K_2[Mg(NH_2)_4]$. Even the sparingly soluble sodium amide dissolves in solutions of rubidium amide to form the complex amide $Rb_2[Na(NH_2)_3]$. Since the amide ion in liquid ammonia is more strongly basic than the hydroxide ion in water, amide complexes are also formed with elements for which no hydroxo complexes are known. Different amido complexes may be formed according to the concentration of the amide ion. For example, two complex amidozirconates have been obtained from zirconium diimide:

$$Zr(NH)_2 + KNH_2 + 2\,NH_3 \longrightarrow K[Zr(NH_2)_5] \xrightarrow{\ KNH_2\ } K_2[Zr(NH_2)_6]$$

The dissolution of metals such as aluminium in aqueous alkali metal hydroxide solutions also has parallels in the ammonia system:

$$Al \xrightarrow{\ KNH_2\ } Al^{3+} + 3e^-_{\ solv.}$$
$$Al^{3+} + 3\,NH_2^- \rightleftharpoons Al(NH_2)_3$$
$$3e^-_{\ solv.} + 3\,NH_3 \rightleftharpoons 3\,NH_2^- + 3/2\,H_2$$
$$Al(NH_2)_3 + NH_2^- \rightleftharpoons [Al(NH_2)_4]^-$$

$$Al + NH_2^- + 3\,NH_3 \rightleftharpoons [Al(NH_2)_4]^- + 3/2\,H_2.$$

The blue colour of solvated electrons in liquid ammonia (cf. Section 2.7) appears during this reaction. This colour disappears with simultaneous formation of hydrogen and amide.

2.4. Solvolysis

Ammonolysis, the solvolysis in liquid ammonia that corresponds to hydrolysis in the aqueous system, is an extremely widely occurring reaction. Since solvolysis depends primarily on the self-ionization of the solvent, and since the degree of self-ionization of liquid ammonia is much smaller than that of water, ammonia has a weaker solvolytic action than water. Solvolysis, can, however, be promoted by the use of higher temperatures (reaction in an autoclave). The primary step in any ammonolysis may be assumed to be the addition of ammonia to the compound in question, frequently with formation of an isolable adduct. This adduct formation is then followed by cleavage of the old bond in the complex and formation of a new bond with the nitrogen.

Thus the alkali metal hydrides react with liquid ammonia to form hydrogen and amides:

$$M-H + NH_3 \xrightarrow{\text{Liq. } NH_3} M-NH_2 + H_2.$$

Some hydrides of group 4 and 5 elements are also ammonolysed in liquid ammonia. For example, when trichlorosilane is slowly heated from −196 °C in contact with liquid ammonia, it forms an imide of silicon:

$$2\,SiHCl_3 + 9\,NH_3 \xrightarrow{\text{Liq. } NH_3} [SiH(NH)]_2NH + 6\,NH_4Cl.$$

When heated, the imide loses ammonia to form polymeric silicon imide, SiNH. Sulphides and selenides also undergo ammonolysis. As was mentioned earlier, carbon disulphide is converted into ammonium dithiocarbamate, while silicon disulphide reacts in a bomb above room termperature to give $Si(NH)_2$. At −33 °C, however, only partial ammonolysis occurs and polymeric products that still contain sulphur bridges are formed. Similarly, the sulphides and selenides of phosphorus and arsenic also give different solvolysis products according to the reaction conditions. The oxides of the heavy metals do not generally react with liquid ammonia. Alkali metal oxides, on the other hand, are completely solvolysed:

$$M_2O + NH_3 \longrightarrow MNH_2 + MOH.$$

Sodium oxide reacts more slowly than the oxides of potassium, rubidium, and caesium. Sodium hydroxide in liquid ammonia is converted in an equilibrium reaction into sodamide and water. Nitrogen oxides have also been ammonolysed. Whereas pure nitrogen(II) oxide, NO, did not react, it was possible, with care, to obtain red nitrosoamine from nitrogen(III) oxide, N_2O_3, and liquid ammonia:

$$N_2O_3 + 2\,NH_3 \longrightarrow NH_4NO_2 + H_2NNO.$$

Nitrosoamine cannot be isolated, since it decomposes readily into nitrogen and ammonium nitrite.

By far most of the ammonolyses carried out so far have been with halogen compounds. Purely ionic halides behave in the same way in liquid ammonia as in water, i.e. they can be recovered unchanged. However, the tendency towards ammonolysis increases with increasing covalent character of the halide.

The ammonolysis of mercury(II) chloride is particularly interesting. The action of liquid ammonia on mercury(II) chloride yields "fusible white precipitate":

$$HgCl_2 + 2\,NH_3 \longrightarrow [H_3N-Hg-NH_3]Cl_2,$$

which is partly converted by reversible solvolysis into"infusible white precipitate", which contains $-Hg-NH_2^+-Hg-NH_2^+$ chains:

$$[Hg(NH_3)_2]Cl_2 \rightleftharpoons [Hg(NH_2)]Cl + NH_4Cl.$$

The equilibrium can be displaced entirely to the right by the addition of a base such as KNH_2, and the final product is a salt of Millon's base:

$$2[Hg(NH_2)]Cl \longrightarrow [Hg_2N]Cl + NH_4Cl.$$

This salt has a three-dimensional network structure consisting of NHg_4 tetrahedra. Mercury(II) bromide and iodide are converted immediately into the salts of Millon's base in liquid ammonia. The reaction of mercury(II) oxide with liquid ammonia yields Millon's base itself:

$$2\,HgO + NH_3 \longrightarrow [Hg_2N]OH \cdot H_2O.$$

Among the group 3 halides, those showing the greatest susceptibility to ammonolysis are the boron trihalides. Boron trichloride forms boron amide:

$$BCl_3 + 6\,NH_3 \longrightarrow B(NH_2)_3 + 3\,NH_4Cl,$$

which can be converted into boron imide above 0 °C and into boron nitride BN at 750 °C. Boron tribromide behaves similarly, while boron triiodide forms boron imide, $B_2(NH)_3$, even at −33 °C. Boron trifluoride is not ammonolysed, but simply forms $BF_3.NH_3$.

Examples of ammonolyses of group 4 halides are those of carbon(IV) iodide, silicon (IV) chloride, disilicon hexachloride, germanium (IV) chloride, tin (IV) chloride, and lead (IV) chloride. The ammonolysis of carbon (IV) iodide yields guanidine. The first step in this reaction is thought to be the formation of iodine cyanide, which reacts further with ammonia to give guanidine:

$$CI_4 + 4\,NH_3 \xrightarrow{\text{fast}} ICN + 3\,NH_4I$$
$$ICN + 3\,NH_3 \xrightarrow{\text{slow}} HN = C(NH_2)_2 + NH_4I.$$

In contrast to carbon(IV) chloride, silicon(IV) chloride is solvolysed even at relatively low temperatures; the only product is silicon diimide, $Si(NH)_2$. This compound is comparable to silicon dioxide, but dissolves only in melting alkali metal amide. It is converted at 900 °C into disilicon dinitride imide, $Si_2N_2(NH)$, and at 1200-1300 °C into silicon nitride. Disilicon hexachloride is ammonolysed at −79 °C to form disilicon triimide $Si_2(NH)_3$. Germanium(IV) chloride behaves in the same manner as silicon(IV chloride, and forms germanium imide:

$$GeCl_4 + 6\,NH_3 \longrightarrow Ge(NH)_2 + 4\,NH_4Cl,$$

which changes at 150 °C into digermanium dinitride imide, $Ge_2N_2(NH)$, and at 300 °C into germanium nitride, Ge_3N_4. The tetrahalides of tin and lead behave differently from those of silicon and germanium. For example, tin tetrachloride gives, not an amide or imide, but an ammono-basic chloride having the composition $SnCl(NH_2)_3$. This compound dissolves in liquid ammonia containing ammonium chloride to form a complex:

$$SnCl(NH_2)_3 + 2\,NH_4Cl \longrightarrow (NH_4)_2[SnCl_3(NH_2)_3].$$

The ammonolysis of lead(IV) chloride or of ammonium hexachloroplumbate, which is easier to handle, leads to a chocolate-brown compound from which lead nitride chloride, $(PbNCl)_6$, can be isolated. Elimination of lead chloride from the latter product yields a substance having the composition $Pb_5N_6Cl_4$, the strained ring system of which is held responsible for its tendency to explode.

Of the halides of group 5 elements, ammonolyses have been carried out in particular with those of phosphorus. A little phosphorus(III) chloride reacts with a large excess of liquid ammonia at -78 °C in the absence of moisture to form mainly phosphorus triamide and ammonium chloride, while when more phosphorus(III) chloride was added to the ammonia, phosphorus amide imide, $HN=P-NH_2$, was also found. Still larger quantities of phosphorus(III) chloride react to form phosphorus nitride, which is amorphous to X rays and insoluble in all solvents:

$$PCl_3 + 4\,NH_3 \longrightarrow PN + 3\,NH_4Cl.$$

The first product of the ammonolysis of phosphorus(V) chloride, according to recent views, is not phosphorus(V) amide; but the triamide of phosphorimidic acid:

$$PCl_5 + 9\,NH_3 \longrightarrow (NH)P(NH_2)_3 + 5\,NH_4Cl.$$

On removal of ammonia, this compound can form phosphorus nitride amide $PN(NH_2)_2$, which exists in the trimeric and tetrameric states, and from which phospham, $PN(NH)$, is finally obtained by heating under vacuum. Arsenic trichloride undergoes ammonolysis in accordance with the following equation:

$$2\,AsCl_3 + 9\,NH_3 \longrightarrow As_2(NH)_3 + 6\,NH_4Cl.$$

Arsenic does not appear to form a triamide. Antimony trichloride was found to form
the imide chloride on ammonolysis:

$$SbCl_3 + 3\,NH_3 \longrightarrow Sb(NH)Cl + 2\,NH_4Cl.$$

The imide chloride could not be isolated by washing with ammonia, which led to the
formation of orange antimony nitride SbN. Owing to the metallic nature of bismuth
its halides are so strongly ionic that they undergo little or no solvolysis in liquid ammonia. Ammonolyses do, however, occur with the halides of the group 5 triad metal.
Thus vanadium(III) chloride forms an amide chloride:

$$VCl_3 + 6\,NH_3 \longrightarrow VCl_2(NH_2) \cdot 4\,NH_3 + NH_4Cl.$$

Vanadium oxide chloride is slowly converted into yellow oxide amide, $VO(NH_2)_3$,
at $-80\,^\circ C$.

The group 6 halides whose behaviour with ammonia has been most extensively studied are the acid halides of sulphur. The reaction of sulphuryl chloride in petroleum
ether with liquid ammonia yields sulphamide and the ammonium salt of imidosul-
phamide:

$$3\,O_2SCl_2 + 12\,NH_3 \longrightarrow O_2S(NH_2)_2 + NH_4[H_2NSO_2\text{-}N\text{-}SO_2NH_2] + 6\,NH_4Cl.$$

When thionyl chloride, $SOCl_2$ is added dropwise to liquid ammonia it produces an
intense red colour, which is due to the formation of the unstable diamide of imido-
disulphurous acid (in the form of the ammonium salt):

$$2\,SOCl_2 + 8\,NH_3 \longrightarrow 4\,NH_4Cl + NH_4[N(SONH_2)_2].$$

Another example from this group is the ammonolysis of tellurium tetrabromide,
which leads to tellurium nitride, a highly explosive, lemon yellow compound:

$$3\,TeBr_4 + 16\,NH_3 \longrightarrow Te_3N_4 + 12\,NH_4Br.$$

Solvolysis reactions have also attained some importance for the preparation of pure
nitrogen bases. For example, hydrazine monosulphate, dioxalate, and diselenate can
be solvolysed with liquid ammonia to form pure hydrazine:

$$N_2H_4 \cdot H_2SO_4 + 2\,NH_3 \longrightarrow (NH_4)_2SO_4 + N_2H_4.$$

The same method has been used for the synthesis of hydroxylamine, methylhydroxy
amine, and semicarbazide. Guanidine cannot be obtained in this way.

The ammonium carbamate formed on dissolution of the ansolvo acid carbon dioxide
in liquid ammonia (cf. Section 2.3.1), which can be used for the preparation of soda

(cf. Section 2.5), can also be used for the production of the industrially important urea. Ammonia and carbon dioxide are liquefied separately and mixed in a ratio of 2:1

$$CO_2 + 2NH_3 \longrightarrow OC(NH_2)_2 + H_2O.$$

The solvolysis is carried out at 153 °C and 100-110 atm, and the intermediate ammonium carbamate is not isolated.

The solvolysis reactions of organic compounds are too numerous to be dealt with in full here, but a few examples will be described to illustrate the preparative value of this type of reaction in organic chemistry. Organic halogen compounds suffer slow ammonolysis at the boiling point of liquid ammonia, the products being primary, secondary, and tertiary amines:

$$RX + 2NH_3 \longrightarrow RNH_2 + NH_4X$$

$$RX + RNH_2 + NH_3 \longrightarrow R_2NH + NH_4X$$

$$RX + R_2NH + NH_3 \longrightarrow R_3N + NH_4X.$$

The principal product in liquid ammonia is the primary amine. Iodides are solvolysed more readily than the other organic halides. Another, very important ammonolysis is the conversion of esters into amides:

$$C_6H_5\text{-}CH=CH\text{-}COOC_2H_5 + NH_3 \longrightarrow C_6H_5\text{-}CH=CH\text{-}CONH_2 + C_2H_5OH.$$
     Ethyl cinnamate   Cinnamic amide

This reaction is catalysed by ammonium ions; the resulting amides can react further under the same conditions to give amidines:

$$CH_3\text{-}CONH_2 + NH_3 \longrightarrow CH_3\text{-}C(NH)NH_2 + H_2O.$$

Ketones can be converted in a similar manner into imides:

$$(C_6H_5)_2CO + NH_3 \longrightarrow (C_6H_5)_2CNH + H_2O.$$
     Acetophenone   Acetophenonimine

2.5. Precipitation Reactions and Complex Formation

Many double decompositions in accordance with the scheme

$$AB + CD \rightleftharpoons AD + CB$$

can be carried out in liquid ammonia, though the products obtained in this solvent are frequently quite different from those found in water because of differences in solubility (cf. Section 2.2.1). The reaction of barium nitrate with silver chloride in liquid ammonia, as was mentioned earlier, leads to barium chloride in accordance with the following equation:

$$Ba(NO_3)_2 + 2\,AgCl \underset{Liq.\;NH_3}{\overset{H_2O}{\rightleftharpoons}} BaCl_2 + 2\,AgNO_3.$$

Silver ions cannot be precipitated from liquid ammonia by chloride, bromide, or iodide ions. Manganese (II) chloride is precipitated from solutions of manganese (II) nitrate in liquid ammonia by the addition of ammonium chloride; ammonium bromide gives a precipitate of manganese (II) bromide. The nitrates of cobalt, nickel, zinc, and the alkaline earth metals react similarly. The chromates and borates of the heavy metals can be obtained by the use of ammonium chromate or borate. Precipitations of salts by sulphates, carbonates, phosphates, or oxalates cannot be carried out in liquid ammonia, since there are no known salts with these anions that are soluble in liquid ammonia.

Double decompositions can be used to obtain substances in the pure state. For example, monomethylarsine is obtained in accordance with the following equation:

$$H_2AsK + ClCH_3 \xrightarrow{-78\,^{\circ}C} KCl\!\downarrow + H_2AsCH_3.$$

Sodium nitrate having a purity of 100% can be prepared as follows:

$$2\,NaCl + Ca(NO_3)_2 + 8\,NH_3 \xrightarrow{+5\,^{\circ}C,\;pressure} CaCl_2 \cdot 8\,NH_3\!\downarrow + 2\,NaNO_3.$$

This double decomposition proceeds quantitatively within 30 minutes. Calcium cyanide, which is difficult to prepare in aqueous solution, can be obtained in liquid ammonia as follows:

$$2\,NH_3 + Ca(NO_3)_2 + 2\,NH_4CN \longrightarrow Ca(CN)_2 \cdot 2\,NH_3 + 2\,NH_4NO_3$$

A reaction of industrial interest is the liberation of cyanamide, which is important in the fertilizer industry, from its calcium salt:

$$CaN\!-\!CN + (NH_4)_2CO_3 \longrightarrow CaCO_3 + H_2N\!-\!CN + 2\,NH_3$$

Owing to the low solubility of ammonium carbonate, this reaction is carried out in suspension. Another precipitation reaction, which is being widely discussed particularly in Japan, is the preparation of sodium carbamate (cf. Section 2.3.1), which could become a key substance for many industrially important products, as is shown in Scheme 1.

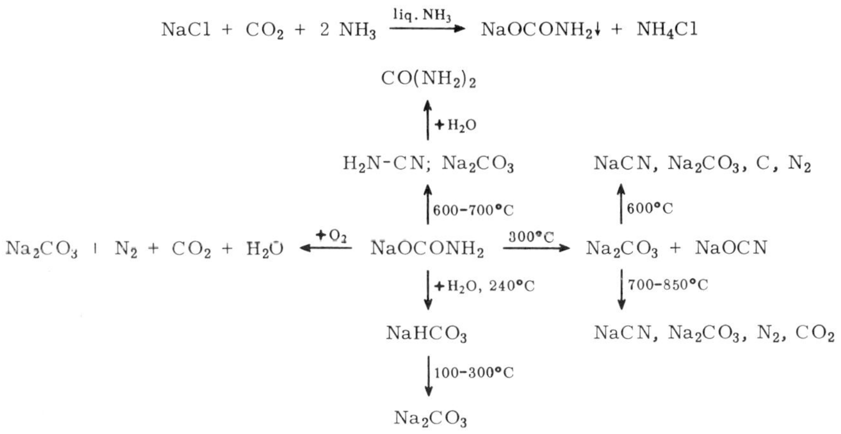

$$NaCl + CO_2 + 2\,NH_3 \xrightarrow{\text{liq. }NH_3} NaOCONH_2{\downarrow} + NH_4Cl$$

$$CO(NH_2)_2$$

$$\Big\uparrow {+H_2O}$$

$$H_2N\text{-}CN;\ Na_2CO_3 \qquad\qquad NaCN,\ Na_2CO_3,\ C,\ N_2$$

$$\Big\uparrow {600\text{-}700°C} \qquad\qquad \Big\uparrow {600°C}$$

$$Na_2CO_3\ |\ N_2 + CO_2 + H_2O \xleftarrow{+O_2} NaOCONH_2 \xrightarrow{300°C} Na_2CO_3 + NaOCN$$

$$\Big\downarrow {+H_2O,\ 240°C} \qquad\qquad \Big\downarrow {700\text{-}850°C}$$

$$NaHCO_3 \qquad\qquad NaCN,\ Na_2CO_3,\ N_2,\ CO_2$$

$$\Big\downarrow {100\text{-}300°C}$$

$$Na_2CO_3$$

Scheme 1. Preparation of sodium carbamate and industrially important products obtainable from it.

Precipitation reactions have also been used in some cases for the identification of solvolysis products. Thus the solvolysis of tetranitrogen tetrasulphide, S_4N_4, in liquid ammonia gives a red solution, from which one mole of lead iodide per mole of tetranitrogen tetrasulphide precipitates the olive green compound $PbS_2N_2 \cdot NH_3$, while two moles of mercury iodide precipitate the bright yellow $HgSN_2 \cdot NH_3$. On the basis of these results, the solvolysis of tetranitrogen tetrasulphide has been formulated as follows:

$$S_4N_4 + 2\,NH_3 \rightleftharpoons S_2N_2H_2 + 2\,S(NH)_2.$$

In addition to the cationic complexes mentioned earlier, in which the ammonia acts as a ligand, amido complexes are also known (cf. Section 2.3.5). Liquid ammonia has also been used simply as a solvent for the preparation of complexes that are sensitive to air and to moisture. Ammonium hexacyanoferrate (II) which can be readily prepared from iron and ammonium cyanide in liquid ammonia:

$$Fe + 2\,NH_4CN \longrightarrow Fe(CN)_2 + H_2{\uparrow} + 2\,NH_3$$

$$Fe(CN)_2 + 4\,NH_4CN \longrightarrow (NH_4)_4[Fe(CN)_6],$$

can also be obtained from aqueous solutions. Ammonium triselenoarsenate (III), on the other hand, can be crystallized as orange-red needles only from liquid ammonia.

$$AeSeNH_2 + 2(NH_4)_2Se \xrightarrow{-78\,°C} (NH_4)_3[AsSe_3] + 2\,NH_3.$$

Complexes e.g. of trivalent titanium can also be readily prepared in liquid ammonia, whereas decomposition occurs in water. The green potassium hexathiocyanatotitana (III)-6-ammonia is formed in accordance with the following equation:

$$Ti_2(SO_4)_3 \cdot 6\,NH_3 + 12\,KSCN + 6\,NH_3 \longrightarrow 2\,K_3[Ti(SCN)_6] \cdot 6\,NH_3 + 3K_2S$$

Complexes containing halide ions as well as amide ions have been found on solvolys of metal halides such as tin (IV) chloride and tungsten (VI) chloride. The solvolysis product of tin tetrachloride is redissolved, with formation of another complex, on addition of ammonium chloride (cf. Section 2.4):

$$SnCl(NH_2)_3 + 2\,NH_4Cl \xrightarrow{-45\,°C} (NH_4)_2[SnCl_3(NH_2)_3].$$

Tungsten hexachloride is solvolysed in boiling ammonia to form tungsten tetrachloride diamide, which forms the complex anion $[WCl_6(NH_2)_2]^{2-}$ in the presence of ammonium chloride:

$$WCl_4(NH_2)_2 + 2\,NH_4Cl \longrightarrow (NH_4)_2[WCl_6(NH_2)_2].$$

Ligand exchange reactions have also been studied in liquid ammonia. Thus cyano and carbonyl ligands in the cyanide and carbonyl complexes of cobalt and nickel can be replaced by carbon monoxide or CN^- ions. In many cases, only addition takes place:

$$2\,K_2[Ni(CN)_3] + 2\,CO \xrightarrow{-40\,°C} 2\,K_2[Ni(CN)_3CO]$$
$$2\,K_2[Ni(CN)_3] + 2\,NO \xrightarrow{-40\,°C} 2\,K_2[Ni(CN)_3NO].$$

Potassium tetracyanoniccolate(0) reacts with carbon monoxide to form potassium dicyanodicarbonylniccolate(0):

$$K_4[Ni(CN)_4] + 2\,CO \xrightarrow{-40\,°C} K_2[Ni(CN)_2(CO)_2] + 2\,KCN.$$

In sodium tetracarbonylcobaltate(−I), on the other hand, a CO molecule can be replaced by a cyanide or thiocyanate ion:

$$Na[Co(CO)_4] + NaCN \longrightarrow Na_2[Co(CO)_3CN] + CO\uparrow$$

$$Na[Co(CO)_4] + NaSCN \longrightarrow Na_2[Co(CO)_3(SCN)] + CO\uparrow.$$

Tetracarbonylnickel dissolves in liquid ammonia with liberation of CO. Compounds having the compositions $Ni(CO)_3NH_3$, $Ni(CO)_2(NH_3)_2$, and $Ni(CO)(NH_3)_3$ are formed. Thus ammonia enters the complex as a ligand in this case. Ligand exchange reactions can also be carried out on hydrates (replacement of water by ammonia). Thus the hydrated heavy metal fluorides $GaF_3 \cdot 3H_2O$, $MnF_2 \cdot H_2O$, $FeF_2 \cdot H_2O$, and $CoF_2 \cdot H_2O$ can be converted into the ammine complexes $GaF_3 \cdot 3\,NH_3$, $MnF_2 \cdot 5\,NH_3$, $FeF_2 \cdot 5\,NH_3$, and $CoF_2 \cdot 5\,NH_3$, which cannot be obtained directly from the anhydrous fluorides. This method can also be used for the dehydration of readily hydrolysable salt hydrates:

$$MgCl_2 \cdot 6\,H_2O + 6\,NH_3 \longrightarrow MgCl_2 \cdot 6\,NH_3 + 6\,H_2O$$

$$MgCl_2 \cdot 6\,NH_3 \xrightarrow{350\,°C} MgCl_2 + 6\,NH_3.$$

It is naturally essential for such dehydrations that no ammonolysis should occur. The replacement of water by ammonia has also played a part in the determination of the constitution of oxide hydrates; in particular it was possible to prepare compounds having definite water and ammonia contents in the cases of the oxide hydrates of silicon and aluminium. For example, on extraction with liquid ammonia, metasilicic acid behaves as follows:

$$6\,SiO_2 \cdot 6\,H_2O + 4\,NH_3 \longrightarrow 6\,SiO_2 \cdot 2\,H_2O \cdot 4\,NH_3 + 4\,H_2O.$$

Lower ammoniates containing three, two, and one molecule of ammonia can be prepared by tensimetric degradation. Metasilicic acid was shown by these reactions to be a homogeneous compound having the formula $6\,SiO_2 \cdot 6\,H_2O$. The products with lower water contents add on solvent molecules in liquid ammonia up to a total number of six, so that they are still metasilicic acids according to their constitution. Extractions with liquid ammonia have also been used with zeolites and permutits, in which water of hydration can be replaced by ammonia, whereas the water belonging to silicic acid remains in the mineral.

2.6. Redox Reactions

Of the many redox reactions that have been carried out in liquid ammonia, only those that do not involve the metals (alkali and alkaline earth metals) that are soluble in liquid ammonia will be discussed here. Redox reactions in which these metals take part will be described in the discussion of the metal solutions (cf. Section 2.7).

Oxidations in the ammonia system are limited by the fact that few oxidizing agents can be used and these are all weak, since stronger oxidizing agents oxidize the solvent. The halogens for example, react with liquid ammonia to give hydrazine and nitrogen via a disproportionation yielding halogenoamines (see below).

On the other hand, liquid ammonia is an excellent solvent for reductions. The only unsuitable reducing agents are those that are particularly strong, such as alkali and alkaline earth metals in the presence of catalysts, since these reduce the solvent to hydrogen and amide.

2.6.1. Oxidations

The oxidizing agents that do not oxidize liquid ammonia to hydrazine or nitrogen include oxygen-rich compounds such as nitrates, potassium permanganate, some oxides of nitrogen, and oxygen itself. Thus alkali metal amides can be oxidized to nitrites by oxygen in liquid ammonia:

$$2\,MNH_2 + 3\,O_2 \longrightarrow 2\,MNO_2 + 2\,H_2O$$
$$2\,MNH_2 + 2\,H_2O \longrightarrow 2\,MOH + 2\,NH_3$$

$$4\,MNH_2 + 3\,O_2 \longrightarrow 2\,MNO_2 + 2\,MOH + 2\,NH_3$$

Under more vigorous conditions, alkali metal amides can be oxidized with potassium nitrate:

$$3\,KNH_2 + 3\,KNO_3 \xrightarrow[+25\,°C]{pressure} 3\,KNO_2 + 3\,KOH + N_2 + NH_3.$$

Still more forcing conditions make it possible to obtain the azides, which are important in industry:

$$3\,KNH_2 + KNO_3 \xrightarrow[110\,-\,140\,°C]{pressure} KN_3 + 3\,KOH + NH_3.$$

Sodium azide can be obtained from sodium nitrate and sodium amide, or lead azide from lead nitrate and potassium amide. Potassium amide is oxidized to nitrogen by potassium permanganate, with formation of potassium manganate(VI):

$$6\,KMnO_4 + 6\,KNH_2 \xrightarrow{-33\,°C} 6\,K_2MnO_4 + N_2 + 4\,NH_3\,.$$

The reaction of nitrous oxide N_2O with alkali metal amides is of industrial interest for the preparation of azides. The reaction with sodium amide (obtained from sodium metal, see Section 2.7) proceeds in accordance with the equation:

$$2\,NaNH_2 + N_2O \longrightarrow NaN_3 + NaOH + NH_3$$

The oxidations in liquid ammonia also include nitriding reactions that result in changes in valency. Apart from a few heavy metal amides, imides, and nitrides (e.g. mercury nitride Hg_3N_2), the most effective nitriding agent is hydrazoic acid HN_3. By analogy with nitric acid in the aqueous system, this acid may be regarded

$$HONO_2 \longrightarrow HONO \longrightarrow (HO)_2NH \longrightarrow HONH_2 \longrightarrow NH_3$$

$$HNNN \longrightarrow H_2NNNH \longrightarrow (H_2N)_2NH \longrightarrow H_2NNH_2 \longrightarrow NH_3$$

| Hydrazoic acid | Triazene | Triazane | Hydrazine | Ammonia |

as "ammono-nitric acid".

An example of these nitriding reactions is the preparation of guanidine from methylamine and ammonium azide:

$$CH_3NH_2 + 3\,NH_4N_3 \xrightarrow[+\,100\,°C]{pressure} HN = C \begin{matrix} \nearrow NH_2 \\ \searrow NH_2 \end{matrix} + 3\,N_2 + 4\,NH_3$$

2.6.2. Reductions

Reductions in which the solvent ammonia is not reduced to hydrogen and amide have been mainly investigated with insoluble metals and alloys, since this is very important in connection with corrosion problems. The metals situated above hydrogen in the electromotive series for liquid ammonia (cf. Section 2.2.4) should be able to liberate hydrogen from ammono acids as follows:

$$M + 2\,NH_4^+ \longrightarrow M^{2+} + H_2 + 2\,NH_3$$

$$NH_4^+ + e^- \longrightarrow NH_3 + 1/2\,H_2\,.$$

However, since the overvoltage for the process in the second equation can be as high

as about 1V (see Section 2.2.4), many base metals such as iron, nickel, cobalt, and
molybdenum are practically insoluble in ammono acids or dissolve only very slowly.
Lead is strongly attacked only by ammonium chloride, but otherwise behaves as a
noble metal. In agreement with the electromotive series, manganese, zinc, and
cadmium liberate hydrogen from ammono acids, while chromium, aluminium, and
gallium which are in fact base metals, are inert, probably because of passivation by
oxide films. Gold, silver, mercury, and tin, as expected, behave as noble metals. Of
the alloys studied, the chrome-nickel and chrome-steels are particularly resistant,
whereas copper-tin and copper-tin-zinc alloys are attacked. Some metals also dissolve
with liberation of hydrogen in ammono-basic solutions. Zinc, cadmium (as amalgams)
and aluminium dissolve under the action of alkali metal amides. It has been found
in some cases that the free alkali metals (as amalgams) are formed as intermediates,
but react further with ammonia under the influence of the metal still present to form
amide and hydrogen. The fact that the electropositive metals sodium and potassium
can be reduced by other metals, though this is not to be expected from the electromotive series, is probably due in many cases to solubility factors. Examples are:

$$Al + 3\,NaCl \rightleftharpoons 3\,Na + AlCl_3$$
$$Be + 2\,NaCl \rightleftharpoons 2\,Na + BeCl_2.$$

Sodium and potassium can be precipitated in the form of their amalgams by amalga-
mated cadmium, zinc, aluminium, or magnesium. The reduction of noble metals as
their salts by less noble metals, as in the aqueous system, naturally also occurs in
liquid ammonia, e.g.

$$Fe + Hg(SCN)_2 \longrightarrow Fe(SCN)_2 + Hg$$
$$Zn + Hg(CN)_2 \longrightarrow Zn(CN)_2 + Hg$$
$$Cu + Cu(CN)_2 \longrightarrow 2\,CuCN.$$

The last of these reactions is an illustration of the relatively high stability of cop-
per(I) compounds in liquid ammonia. Copper and copper(II) sulphate also react to
an equilibrium with copper(I) sulphate.

2.6.3. Disproportionations

A type of redox reaction that is known from the aqueous system and is also observed
in liquid ammonia is disproportionation. Whereas chlorine, for example, disproportionates in water only to an equilibrium with hydrochloric acid and hypochlorous acid:

$$Cl_2 + 2\,H_2O \rightleftharpoons ClOH + H_3O^+ + Cl^-,$$

the reaction in liquid ammonia proceeds to completion with formation of monochloro-amine and ammonium chloride:

$$Cl_2 + 2\,NH_3 \longrightarrow ClNH_2 + NH_4^+ + Cl^-.$$

Monochloroamine, which is of interest for the synthesis of hydrazine, can be obtained in yields of 50 to 60% by introduction of chlorine diluted with nitrogen, after cooling to $-78\,°C$, into liquid ammonia at the same temperature. If excess chlorine is used, nitrogen trichloride is formed:

$$4\,NH_3 + 3\,Cl_2 \longrightarrow 3\,NH_4Cl + NCl_3.$$

Monochloroamine, which is relatively stable in liquid ammonia below $-75\,°C$, reacts with ammonia to form hydrazine, as is known from the Raschig synthesis in the aqueous system:

$$NH_2Cl + 2\,NH_3 \longrightarrow N_2H_4 + NH_4Cl$$

The hydrazine formed is further oxidized (particularly in ammono-acidic or ammono-basic media) by monochloroamine:

$$N_2H_4 + 2\,NH_2Cl \longrightarrow N_2 + 2\,NH_4Cl.$$

In the neutral pH range hydrazine can be obtained in 80% yield. Monochloroamine decomposes in ammono-basic media to form nitrogen and ammonia:

$$3\,NH_2Cl + 3\,KNH_2 \longrightarrow N_2 + 4\,NH_3 + 3\,KCl.$$

Bromine disproportionates in liquid ammonia in the same way as chlorine:

$$Br_2 + 2\,NH_3 \xrightarrow{\text{melting } NH_3\,^{*)}} NH_2Br + NH_4Br.$$

Solutions of monobromoamine in liquid ammonia or ether are colourless to pale yellow, but become intensely red-violet on freezing with liquid nitrogen, owing to the possibility of delocalization of electrons in the solid state; the colour disappears when the solution melts. Black-violet monobromoamine can be precipitated from solutions in ether by cold pentane.

Iodine reacts via $I_2\,2NH_3$ (which corresponds to the halogen hydrates $Cl_2.8H_2O$

---

*) Consist of solid and liquid NH_3. The heat of reaction is absorbed as heat of fusion.

and $Br_2 \cdot 10H_2O$) to form the green 3-ammine of nitrogen triiodide, which is practically insoluble in liquid ammonia:

$$3\,I_2 + 16\,NH_3 \rightleftharpoons 3\,NH_4I \cdot 4\,NH_3 + NI_3 \cdot 3\,NH_3.$$

It can be converted under vacuum at $-75\ ^\circ C$ into the red-brown NI_3NH_3, in which solvated NI_4 tetrahedra are joined by common iodine atoms to form chain macromolecules. The products with lower ammonia contents are very explosive. When iodine reacts with a large excess of melting ammonia, the disproportionation leads, not to the green compound, but to the red monoiodoamineammonia, which is also practically insoluble in liquid ammonia:

$$3\,I_2 + 9\,NH_3 \xrightarrow{\text{melting } NH_3} 3\,(NH_2I \cdot NH_3)\!\downarrow + 3\,NH_4I.$$

and from which black monoidoamine NH_2I can be obtained.

Other examples of disproportionations are the reactions of sulphur and of some carbonyls with liquid ammonia. Sulphur disproportionates to an equilibrium with ammonium hydrogen sulphide and tetranitrogen tetrasulphide, with appearance of a purple colour:

$$10\,S + 10\,NH_3 \rightleftharpoons 6\,NH_4HS + S_4N_4.$$

Dimeric and tetrameric carbonylcobalt react as follows at $-33\ ^\circ C$:

$$3[Co^{\,o}(CO)_4]_2 + 12\,NH_3 \longrightarrow 2[Co^{II}(NH_3)_6][Co^{-I}(CO)_4]_2 + 8\,CO$$

$$3[Co^{\,o}(CO)_3]_4 + 24\,NH_3 \longrightarrow 4[Co^{II}(NH_3)_6][Co^{-I}(CO)_4]_2 + 4\,CO.$$

A reaction very similar to the disproportionation in water is observed with potassium manganate(VI) in ammono-acidic solution:

$$3\,K_2\,Mn^{VI}O_4 + 4\,NH_4Cl \xrightarrow{-73\ ^\circ C} 2\,KMn^{VII}O_4 + Mn^{IV}O_2 + 4\,KCl$$
$$+ 4\,NH_3 + 2\,H_2O.$$

2.7. Solutions of Alkali and Alkaline Earth Metals in Liquid Ammonia

2.7.1. Physico-Chemical Properties and Structure

One of the most remarkable phenomena in the liquid ammonia system is found in solutions of alkali and alkaline earth metals, which are deep blue at low concentrations and bronze at higher concentrations. Since their discovery by Weyl (1863) these solutions have been the subject of many physical and chemical investigations.

In addition to the alkali and alkaline earth metals, europium and ytterbium are also soluble in liquid ammonia. Extremely unstable blue solutions of aluminium have also been obtained by cathodic reduction of aluminium triiodide in liquid ammonia. The cathodic reduction of tetramethylammonium iodide, but not of ammonium iodide, also leads to blue solutions which are stable for several hours. Alkali metals also give blue solutions in a number of other solvents. These include some amines and ethers, such as ethylene glycol dimethyl ether, $CH_3OCH_2-CH_2OCH_3$, alcohols, and even water (though the solutions in this case are extremely unstable), as well as fused halides, hydroxides, and amides.

All solutions of metals in liquid ammonia are metastable, though they can be stored for long periods in the absence of catalysts (impurities). Catalysts, and in particular finely divided metals (platinum asbestos, platinum sponge, Raney nickel), favour decomposition in accordance with

$$M + x\,NH_3 \longrightarrow M(NH_2)_x + \frac{x}{2}H_2.$$

This decomposition is used for the preparation of alkali and alkaline earth metal amides (amide reaction). It corresponds to the reaction of alkali metals with water. The catalytic activity of many metal salts (particularly iron salts) is due to the fact that the salt is reduced first, and the resulting finely divided metal catalyses the amide formation.

Just as remarkable as the fact that the metals are soluble is the extent of their solubility. Thus 100 g of liquid ammonia at $-33\,^\circ C$ dissolve 10.9 g of lithium, 24.8 g of sodium, or 46.4 g of potassium. The saturation concentration of caesium at $-50\,^\circ C$ is reported to be 334 g per 100 g of ammonia. 33.6 g of calcium dissolve per 100 g of ammonia at $-35\,^\circ C$.

Solid bronze-coloured ammonia adducts of the type $Ca(NH_3)_6$, $Sr(NH_3)_6$ and $Ba(NH_3)_6$ can be isolated from the solutions of the alkaline earth metals, whereas the detection of adducts of the alkali metals is very uncertain. Phase separation has been observed in a solution of sodium in liquid ammonia ($4-5$ atom-% of sodium) below $-41.7\,^\circ C$, with formation of a less dense bronze coloured layer having a higher sodium content and a clear deep blue layer. Similar phase separations are observed in solutions of potassium and the alkaline earth metals, whereas no phase separation can be detected in the case of caesium.

The densities of the metal solutions decrease with increasing metal concentration. A saturated lithium solution (approx. 3.7 mole of NH_3 per g-atom of lithium) has a density of 0.477 g/ml at 19 °C, and is thus the lightest known liquid at this temperature; the value found for 29.2 moles of ammonia per g-atom of lithium at $-33\,^\circ C$ was 0.639 g/ml. Thus the quantity of matter per ml of solution decreases with increasing metal concentration. This observation suggests the presence of empty cavities in the solution, and is one of the pieces of experimental evidence in favour of the cavity model discussed below.

Electrical conductivity measurements have also led to very interesting observations. All metal solutions have a very high conductivity over a wide concentration range, this conductivity being of the same order in highly concentrated solutions as that of metallic conductors. Thus the specific conductivity of a saturated sodium solution at -33.5 °C is $5047\Omega^{-1}cm^{-1}$, while that of metallic mercury at $+20$ °C is $10600\Omega^{-1}cm^{-1}$. The equivalent conductivities follow the same course for all metal solutions; with increasing dilution, they initially decrease, pass through a minimum at about 0.05 mole/l, and increase asymptotically toward a limiting value with further dilution (Fig. 3).

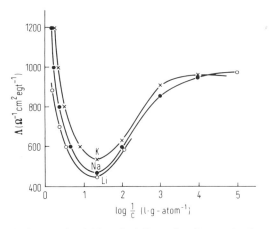

Fig. 3. Equivalent conductivities of solutions of sodium, potassium and lithium in liquid ammonia at -33.5 °C (1/c = litres of pure NH_3 in which 1 g-atom of metal is dissolved.)

Further information about the nature of the metal solutions has been provided by magnetic measurements. Very concentrated solutions exhibit the weak paramagnetism of pure solid metals. At medium concentrations the sample becomes diamagnetic, and paramagnetism appears at high dilutions, the magnitude of this paramagnetism corresponding to the presence of particles having a magnetic moment of 1 Bohr magneton. It is therefore concluded that unpaired electrons are present only in very dilute solutions.

The blue colour of these solutions is due to a visible part of a strong absorption band having a maximum in the infrared range at about 15000 Å (Fig. 4).

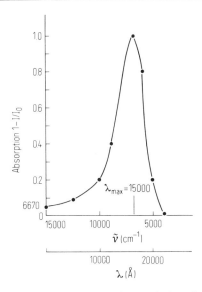

Fig. 4. Absorption spectrum of a dilute solution of potassium in liquid ammonia.

The fact that the absorption maximum is independent of the nature of the dissolved metal suggests that the light is absorbed by electrons released from the metals (see below).

Regarding the structure of these solutions, all the models that are currently being considered are based on Kraus's view that alkali and alkaline earth metals in liquid ammonia dissociate into solvated ions and electrons:

$$Na + (x + y) NH_3 \longrightarrow [Na(NH_3)_x]^+ + [e(NH_3)_y]^-.$$

This basic idea has led to the development of the cavity model for very dilute solutions, in which the isolated electron is assumed to occupy a cavity surrounded by solvent molecules. Numerous calculations of the energy of the electrons in the cavity, the dimensions of the cavity, etc. have been carried out on the basis of this theory. The radius of a cavity is now thought to be $3 - 4$ Å. This value has recently been confirmed experimentally, and is also reflected in the low density of the metal solutions. In the cavity, the electron is situated in the lowest energy level (1s state). It is assumed to move in the orbitals of the protons of the ammonia molecules surrounding the cavity. The absorption spectrum of a metal solution is explained by the assumption that the electron passes from the 1s into the 2p state.

Another assumption, which is particularly important for the interpretation of the magnetism and the electric conductivity, is that a cavity can also be occupied by two electrons having opposite spins. This leads to the following equilibrium:

$$
2\,e^-\text{(cavity)} \quad
\begin{array}{c}
\text{falling temp.}\\
\text{increasing concn.}\\
\rightleftharpoons\\
\text{decreasing concn.}\\
\text{rising temp.}
\end{array}
\quad e_2^{2-}\text{(cavity)}
$$

paramagnetic   diamagnetic

In particular, the transition from paramagnetism to diamagnetism on moderate concentration of dilute metal solutions can be readily explained by the displacement of the equilibrium from left to right. However, additional assumptions are required in order to describe correctly the weak paramagnetism of more concentrated solutions. By analogy with views on the self-localization of electrons in crystals, Russian workers have developed the polaron theory, a polaron being the name given to the stationary state resulting from the polarization of the dielectric ammonia by an electron. The theory also predicts the existence of F_1 and F_2 centres in which single electrons and electron pairs occur in surroundings different from those in the polaron. All three states of the electron are assumed to exist in a concentration-dependent and temperature dependent equilibrium with one another. The polaron theory is related to the cavity model, but this theory can also be used to explain the phenomena in concentrated solutions, for which the cavity model fails.

Both models break down when applied to the bronze-coloured concentrated metal solutions. This failure led to the development of the expanded metal model, which is also known as the cluster theory. In view of the isolation of 6-ammonia adducts of alkaline earth metals, it is assumed that complexes of one metal cation and about six ammonia molecules (cluster) are formed in solution, the protons of the ammonia molecules being directed outward. The electron of the metal moves in the proton orbitals of the ammonia molecules, and so describes an expanded path about the residual metal ion (expanded metal). The solution is also assumed to contain solvated electrons and ions formed by the dissociation of the complex. When the concentration is increased, monomeric complexes can combine to form dimeric complexes and higher associates:

$$
2M_{solv}^+ \;+\; 2e_{solv}^- \;\rightleftharpoons\; 2\,[M^+(NH_3)_6 e^-]
\qquad
\begin{array}{c}
\text{increasing concn.}\\
\rightleftharpoons\\
\text{decreasing concn.}
\end{array}
$$

paramag-   monomer paramagnetic   netic

$$[M^+(NH_3)_6 e^-]_2 \quad \underset{\text{decreasing concn.}}{\overset{\text{increasing concn.}}{\rightleftharpoons}} \quad \frac{2}{x}[M^+(NH_3)_6 e^-]_x$$

dimer   polymer

diamagnetic

Thus in the concentrated metal solutions the solvated electron and the monomer are paramagnetic while the other states are diamagnetic. The polymer has an electical conductivity of the same order of magnitude as that of pure metals.

Though solutions of metals in liquid ammonia have been the subject of intense research for more than a century, satisfactory quantitative explanations cannot yet be given for all the phenomena. The models mentioned provide reliable interpretations only for certain concentration ranges. It is therefore impossible to decide in favour of any one of these models, and a combination of models appears to be the most reasonable solution at present.

2.7.2. Reactions of Metals Dissolved in Liquid Ammonia

Solutions of alkaline earth metals in liquid ammonia are very strong reducing agents. Lithium, sodium, potassium, rubidium, caesium, magnesium, calcium, strontium, barium, europium, and ytterbium reduce water to hydrogen, and so cannot be used as reducing agents in water. This type of reduction has consequently contributed greatly to the use of liquid ammonia as a solvent. Moreover, despite the low temperatures, many of these reactions proceed rapidly. It is therefore generally unnecessary to operate at high temperatures under pressure, and this partly offsets the extra trouble and expense of cooling. Since these reductions are subject to interference by the amide reaction (Section 2.7.1) the system must be kept free of any substances that catalyse this reaction.

In redox reactions with solutions of metals in liquid ammonia, electrons are transferred to an atom, a group of atoms, or an ion, the metal cation generally having no effect on the reaction. An exception to this rule is the reduction of silver thiocyanate, which yields different products according to whether sodium, potassium, or calcium is used:

$$AgSCN + Na \longrightarrow Ag + NaSCN$$

$$AgSCN + K \longrightarrow Ag + KSCN$$

$$AgSCN + 3K \longrightarrow Ag + KCN + K_2S$$

$$2\,AgSCN + 2\,Ca \longrightarrow 2\,Ag + Ca(CN)_2 + CaS.$$

Reactions of solutions of alkali and alkaline earth metals in liquid ammonia have been carried out with many elements. Among the reaction products of the metals of groups 3 and 4, special mention should be made of the polyanionic compounds. These are generally prepared from the halides of the metals, which are immediately reduced by the alkali metals to give finely divided metals, which react with further alkali metals:

$$MHal_n + n M^I \longrightarrow n M^I Hal + M$$
$$M + n M^I \longrightarrow M^{n-} + n M^{I+}$$
$$M^{n-} + m M \longrightarrow M^{n-}_{m+1}.$$

The metal anions formed are stabilized by the addition of further metal atoms. This yields compounds of the type Na_4Pb_9 and Na_4Sn_9, some of which are soluble, and can also be obtained by extraction of sodium-lead or sodium-tin alloys with liquid ammonia. The salt-like nature of these polyanionic salts in liquid ammonia is shown by their electrical conductivity, by the anodic deposition of lead on electrolysis of Na_4Pb_9, and by chemical reactions:

$$Na_4Pb_9 + 2 Pb^{2+} \longrightarrow 4 Na^+ + 11 Pb\downarrow$$
$$Na_4Pb_9 + 4 NH_4^+ \longrightarrow 9 Pb + 2 H_2 + 4 NH_3 + 4 Na^+$$
$$K_4Pb_9 + 2 Ca \longrightarrow Ca_2Pb_9\downarrow + 4 K.$$

In many cases the metal that forms the complex anion can be replaced by another metal.

$$K_4Pb_9 + 8 Sn \longrightarrow K_4Sn_8 + 9 Pb\downarrow.$$

On removal of ammonia, the polyanionic salts are converted into alloys. Compounds having the formulae Na_3As, Na_3As_7, Na_3Sb, Na_3Sb_7, and Na_3Bi_3 have been prepared from the elements of group 5.

Many compounds, some of which are polyanionic in nature, have also been obtained with nonmetals. Some examples are given in Table 6. Graphite and silicon do not form polyanionic compounds.

Graphite gives salt-like, insoluble interstitial compounds containing ammonia as well as the alkali metal, such as the deep blue preparations having the ideal formula $C_{12}M(NH_3)_2$. The additional ammonia present also stabilizes the otherwise unknown compounds of graphite with lithium and sodium, and a europium-graphite compound has even been isolated from liquid ammonia.

The alkali metal-oxygen compound formed depends on the reaction conditions (excess or deficiency of metal, temperature, strength of the oxygen current). When

Table 6. Products formed on reduction of non-metals with solutions of metals in liquid ammonia

Non-metal	Reducing metal	Products	Non-metal	Reducing metal	Products
P	Li	Li_4P_2	S	Li	$Li_2S, Li_2S_2, Li_2S_4, Li_2S_x$
	Na	$Na_2H_2P_2, Na_4P_2 \cdot 2\,NH_3$		Na	$Na_2S, Na_2S_2, Na_2S_4, Na_2S_x$
	K	K_3P, K_6P_4		K	$K_2S, K_2S_2, K_2S_4, K_2S_x$
O_2	Li	Li_2O, Li_2O_2, LiO_2		Rb, Cs	Rb_2S, Cs_2S
	Na	Na_2O, Na_2O_2, NaO_2 (unstable)		Ca	CaS, CaS_2
	K	KO_2, K_2O_2	Se	Li	Li_2Se, Li_2Se_2
	Rb	Rb_2O_2, RbO_2		Na	$Na_2Se, Na_2Se_2, Na_2S_3,$ $Na_2Se_4, Na_2Se_5, Na_2Se_6$
	Cs	Cs_2O_2, Cs_2O_3, CsO_2		K	$K_2Se, K_2Se_2, K_2Se_3, K_2Se_4,$ K_2Se_5
	Mg	MgO_2		Rb, Cs	Rb_2Se, Cs_2Se
	Ca	CaO, CaO_2	Te	Li	Li_2Te
	Sr	SrO, SrO_2		Na	$Na_2Te, Na_2Te_2, Na_2Te_3,$ Na_2Te_4
O_3	Ba	BaO, BaO_2		K	K_2Te, K_2Te_3
	Li	$LiO_3 \cdot 4\,NH_3$		Rb, Cs	Rb_2Te, Cs_2Te

oxygen is passed rapidly through solutions of potassium, rubidium, and caesium at $-60\ ^{\circ}C$, the yellow hyperoxides KO_2, RbO_2 and CsO_2 are formed in good yields. The peroxides M_2O_2 can also be obtained as intermediates. In the case of lithium, the hyperoxide can be obtained only by rapid oxidation at $-78\ ^{\circ}C$. This oxide decomposes even at $-33\ ^{\circ}C$ into two oxides, Li_2O and Li_2O_2. Small quantities of hydroxide, nitrite, and nitrate (resulting from the oxidation of amide intermediates) are formed as by-products in the oxidation of the metals.

Like insoluble metals (e.g. zinc or iron), the solutions of alkali and alkaline earth metals react with ammono acids to liberate hydrogen:

$$M + n\,NH_4X \longrightarrow MX_n + n\,NH_3 + \frac{n}{2}\,H_2.$$

This reaction can be used to remove excess metal from a reaction mixture. The ammonium ions formed during ammonolysis can also be detected by the evolution of hydrogen on addition of alkali metals. Attempts to prepare ammonium radicals by this reaction have been unsuccessful and always resulted in the evolution of hydrogen. The results obtained by the use of sodium amalgam are also inconclusive; it is more than uncertain whether ammonium amalgam is formed, particularly since hydrogen is liberated above $-30\ ^{\circ}C$.

The reactions of germanium tetrahydride, tin tetrahydride (stannane), hydrazine, phosphine, and arsine (which are weakly acidic in liquid ammonia) with the metals that are soluble in liquid ammonia lead to the liberation of hydrogen and formation of salts having the formulae $Na[GeH_3]$, $K[GeH_3]$, $Na[SnH_3]$, Na_2SnH_2, $Na[N_2H_3]$, $Na[PH_2]$, $K[PH_2]$, $Na[AsH_2]$ or $K[AsH_2]$.

Mention should also be made of the reactions with oxides of nonmetals, such as carbon monoxide, carbon dioxide, nitrous oxide, and nitric oxide. Carbon monoxide reacts with solutions of alkali and alkaline earth metals to give the so-called alkali and alkaline earth metal carbonyls having the formulae LiCO, NaCO, KCO, and $Ca(CO)_2$, which are very unstable to oxygen and atmospheric moisture. X-ray studies indicate that the yellow "carbonylpotassium" is in fact potassium acetylenediolate $KOC \equiv COK$. The reaction with carbon dioxide leads first to ammonium carbamate, which reacts as an ammono acid with the metal solutions to liberate hydrogen:

$$NH_4OC(O)NH_2 + M^I \longrightarrow M^IOC(O)NH_2 + 1/2\,H_2.$$

The resulting metal carbamate can be reduced by the hydrogen to metal formate:

$$M^IOC(O)NH_2 + H_2 \xrightarrow{\ >-50\ ^{\circ}C\ } M^IOC(O)H + NH_3.$$

The reaction of the metal solutions with nitrous oxide is of interest in connection with the azide synthesis (cf. Section 2.6).

$$2 N_2O + 4 K + 2 NH_3 \longrightarrow 2 KNH_2 + 2 N_2 + 2 KOH$$
$$2 KNH_2 + N_2O \longrightarrow KN_3 + KOH + NH_3 .$$

In an industrially interesting variant, nitrous oxide is saved by the addition of catalysts for the formation of amide (water, rust, or platinum black) to the reaction mixture:

$$2 Na + N_2O + NH_3 \xrightarrow{\text{Catalyst}} NaN_3 + NaOH + H_2 .$$

Nitric oxide gives products having the formula $M^I NO$, which do not exhibit the reactions of trans-hyponitrites, and may be cis-hyponitrites.

Heavy metal sulphides and selenides having a layer lattice e.g. molybdenum(IV) sulphide, tungsten(IV) selenide, and titanium(IV) sulphide, react with the metal solutions in the same way as graphite with incorporation of metal (and in some cases ammonia). Examples of products observed are $MoS_2Li_{0.8}(NH_3)_{0.8}$, $WSe_2K_{0.4}$, and $TiS_2Na_{0.8}$.

The reactions of salts of the group 8 triad metals with alkali metals in liquid ammonia are of some importance, since they lead to very finely divided, catalytically active metal precipitated, e.g.:

$$CoI_3 + 3 K \longrightarrow Co + 3 KI .$$

The problem of using highly active metal precipitates obtained in this way as hydrogenation catalysts is that they are generally also good catalysts for amide before the heavy metal salt has been quantitatively reduced. The metal precipitates consequently generally contain only 90% of metal together with heavy metal hydroxide, alkali metal salts, and absorbed hydrogen. Nevertheless, some of them are as effective as Raney nickel.

The metal solutions are also of interest for the preparation of unusual valency states of transition metals in cyanide complexes. For example, the cyano complexes of monovalent and of zero-valent nickel ($K_2[Ni(CN)_3]$ and $K_4[Ni(CN)_4]$) can be obtained by reduction of potassium tetracyanoniccolate(II), $K_2[Ni(CN)_4]$, with potassium in liquid ammonia. The red $K_2[Ni(CN)_3]$ must be formulated as a dinuclear complex $K_4[Ni_2(CN)_6]$. The reaction of the cobalt complex $K_3[Co(CN)_6]$ with less than one equivalent of potassium yields the compound $K_3[Co^I(CN)_4]$, while further potassium gives the complex $K_8[Co^0(CN)_4]_2$, which is a brown-violet pyrophoric substance isoelectronic with $[Co(CO)_4]_2$. Manganese also gives complexes

containing monovalent and zero-valent metal, such as the complex
$K_5[Mn^I(CN)_6] \cdot NH_3$.

The metal solutions react with carbonylmetals to form the alkali metal salts of the carbonylhydrogens. In an earlier preparation, the ammonium salts of the carbonyls were used as the starting materials:

$$(NH_4)_2[Fe(CO)_4] + 2\,Na \longrightarrow Na_2[Fe(CO)_4] + H_2 + 2\,NH_3.$$

The method used at present is a direct reaction of the monomeric or polynuclear metal carbonyls with the metal solutions. For example, all three carbonylirons $(Fe(CO)_5, Fe_2(CO)_9, [Fe(CO)_4]_3)$ lose carbon monoxide and form the salt $Na_2[Fe(CO)_4]$. The following carbonyls are also converted into the corresponding carbonylmetallates by sodium solutions: $[Co(CO)_4]_2$ into $Na[Co(CO)_4]$, $W(CO)_6$ and $Mo(CO)_6$ into $Na_2[W(CO)_5]$ and $Na_2[Mo(CO)_5]$, and $Cr(CO)_6$ into $Na_2[Cr(CO)$
The reduction of organic substances is of great preparative importance, as well as of theoretical interest. Bonds are broken with reversible addition of one or two electrons to form radicals and anions, which are subsequently protonated:

$$X-Y \begin{cases} \underset{-e}{\overset{+e}{\rightleftharpoons}} X\cdot + Y^- \underset{-X\cdot,\,H^+}{\overset{+X\cdot,\,H^+}{\rightleftharpoons}} X-X + HY \\[2mm] \underset{-2e}{\overset{+2e}{\rightleftharpoons}} X^- + Y^- \underset{+2\,H^+}{\overset{+2H^+}{\rightleftharpoons}} HX + HY \end{cases}$$

The reduction of double bonds ($X=Y$) can be represented by a similar scheme. This also leads to reversible addition of electrons and formation of ionic species of the types $X-Y^-$ and $X-Y^-$, which are then converted into saturated products by dimerization or protonation. The electrons present in the metal solutions possess the energy required for addition. The formation of the anions and anion-radicals is favoured by the low acidity and the dipole and ionization forces of ammonia. The proton required for this last step comes from the ammonia, though it may also be supplied by acids in ammonia. Acids thus accelerate the reduction or even give entirely different products. For example, benzene is reduced to 1,4-dihydrobenzene by sodium in liquid ammonia in the presence of alcohol, but not in its absence.
Since protonation in the absence of acids such as alcohols or ammonium salts leads to an increase in the amide ion concentration, and hence to an increase in the basicity of the medium, base-catalysed rearrangements are also possible. Thus sodium reacts with naphthalene, to give 1,4-dihydronaphthalene; this is rearranged to 1,2-

dihydronaphthalene by the amide ions formed, and is finally reduced to tetralin. In the presence of alcohol, on the other hand, only 1,4-dihydronaphthalene is formed.

There are innumerable examples of the reductive cleavage of both carbon-carbon and carbon-heteroatom bonds. For example, hexaphenylethane is converted into sodium triphenylmethanide:

$$(C_6H_5)_3C-C(C_6H_5)_3 + 2\,e \longrightarrow 2\,(C_6H_5)_3C^-$$

The cleavage of an oxygen-carbon bond is observed e.g. in allyl alcohol.

$$CH_2=CH-CH_2OH + 2\,e \longrightarrow OH^- + \overset{\frown}{CH_2 \cdots CH \cdots CH_2} \overset{H^+}{\longrightarrow}$$
$$OH^- + CH_3-CH=CH_2$$

Cleavages of this kind have also been observed in some allyl and benzyl ethers, whereas normal ethers are inert. The cleavage of esters RCO_2R' is more complicated, and depending on the quantity of sodium used, leads to diketones RCO-COR, acyloins RCO-CHOH-R and other products. Nitrogen-carbon bonds can also be broken. This is illustrated by the conversion of triphenylmethylamine into sodium triphenyl-methanide and amide:

$$(C_6H_5)_3CNH_2 + 2\,e \longrightarrow (C_6H_5)_3C^- + NH_2^-.$$

The cleavage of halogen-carbon bonds must be mentioned because of its analytical importance. This reaction applies to all the halogens, including fluorine, irrespective of whether the halogen is attached to saturated or unsaturated carbon atoms. In the case of monohalogenated hydrocarbons the reaction leads almost exclusively to the saturated hydrocarbons, together with small quantities of primary amines. Dihalogeno compounds also give olefins (e.g. the dihalide $CH_3-CHBr-CH_2Br$ gives the olefin $CH_3-CH=CH_2$).

π-Electron systems (double bonds, aromatic systems) are generally reduced only with difficulty or not at all by pure metal solutions. However, many olefins (e.g. 1-hexene) can be reduced with methanol and sodium in liquid ammonia at −33 °C (2-hexene is not attacked under these conditions). Examples of stronger reducing agents for olefins are solutions of lithium in methylamine, ethylamine, propylamine, or ethylenediamine, since these can be used at higher temperatures, Acetylenes can be reduced to olefins (though only incompletely) even in the absence of proton donors (e.g. acetylene gives ethylene and propyne gives propylene). The incompleteness of the reaction is due to a side reaction leading to the formation of acetylides, which cannot accept further electrons because of their negative charge. 1-Alkynes are therefore best reduced to the 1-alkenes in solutions containing ammonium sulphate. If an excess of sodium is avoided, practically no alkanes are formed in this reaction, which is therefore preferable to catalytic hydrogenation, which nearly always yields alkanes that are difficult to separate.

Organometallic compounds have also been reduced with solutions of metals in liquid ammonia. For example, ethylmercuric chloride reacts with sodium in liquid ammonia as follows:

$$C_2H_5HgCl + 3\,Na + NH_3 \longrightarrow NaCl + NaHg + C_2H_6 + NaNH_2.$$

Polymeric dimethyltin reacts with sodium to form red solutions, from which disodium dimethylstannide can be isolated on addition of further sodium

$$2\,(CH_3)_2Sn + 2\,Na \longrightarrow [(CH_3)_2Sn-Sn(CH_3)_2]^{2-} + 2\,Na^+$$

$$[(CH_3)_2Sn-Sn(CH_3)_2]^{2-} + 2\,Na \longrightarrow 2\,[(CH_3)_2Sn]^{2-} + 2\,Na^+.$$

The reaction of triphenylstannane $(C_6H_5)_3SnH$ with sodium in liquid ammonia yields sodium triphenylstannide $(C_6H_5)_3SnNa$, together with small yields of disodium diphenylstannide $(C_6H_5)_2SnNa_2$. Tetraphenylstannane also gives a mixture of sodium triphenylstannide and disodium diphenylstannide. The reaction of alkyl- and aryltin halides with sodium normally results in the cleavage of the halogen-tin bond. For example, trimethyltin chloride gives sodium trimethylstannide. With one mole of sodium per mole of halide, however, the reaction leads first to high yields of hexamethyldistannane, which is then cleaved by further sodium:

$$(CH_3)_3SnSn(CH_3)_3 + 2\,Na \longrightarrow 2\,(CH_3)_3SnNa.$$

Tin-carbon, tin-silicon, and tin-oxygen bonds are broken in a similar manner.

2.8. Closing Remarks on the Ammonia System

The great similarity between liquid ammonia and water and the reactions that can be carried out in the two solvents (acidbase reactions, neutralizations, solvolyses, amphoteric reactions, precipitations, and complex formation reactions) led Franklin, in the first quarter of the century to compare the nitrogen compounds derived from ammonia, which he described as the "nitrogen system of compounds", with the oxygen compounds derived from water. Though the analogy is incomplete, a systematic comparison of the two systems of compounds is nevertheless justifiable and useful. As was shown, the ammonium ions NH_4^+ and the amide ions NH_2^- of liquid ammonia correspond to the hydronium ions H_3O^+ and the hydroxide ions OH^- of water. The ammine complexes have their parallel in the aquo complexes, and the mixed aquo-ammine complexes, which are fairly common, may be regarded as belonging to both systems:

$OC{<}^{OH}_{OH}$	$OC{<}^{OH}_{NH_2}$	$OC{<}^{NH_2}_{NH_2}$	$HNC{<}^{NH_2}_{NH_2}$
Carbonic acid	Carbamide	Urea	Guanidine
$O_2S{<}^{OH}_{OH}$	$O_2S{<}^{OH}_{NH_2}$	$O_2S{<}^{NH_2}_{NH_2}$	
Sulphuric acid	Amidosulphonic acid	Sulphamide	
HOOH	$HONH_2$		H_2NNH_2
Hydrogen peroxide	Hydroxylamine		Hydrazin
$HO(CH_2)_2OH$	$HO(CH_2)_2NH_2$		$H_2N(CH_2)_2NH_2$
Ethylene glycol	2-Aminoethanol		Ethylenediamine

The primary amines correspond to the alcohols; tertiary or even secondary amines may be compared to the ethers, and chloroamine NH_2Cl corresponds to hypochlorous acid $HOCl$.

Liquid ammonia is the only ionizing solvent with strong basicity of any importance. The organic amines and hydroxylamines, such as the compounds $HOCH_2CH_2NH_2$ and $(HOCH_2CH_2)_2NH$, are more limited in their application as solvents and are therefore less important (cf. Section 10). Hydrazine, the basicity of which is slightly less than ammonia, has acquired a certain importance. Hydrazine is liquid from 2 to 113.5 °C. Its dielectric constant at 25 °C is 51.7, and is therefore higher than that of ammonia. It has the disadvantage of possessing strong reducing properties, which lead to deposition of free metals from many metal salts. Hydrazine is readily oxidized to nitrogen in the presence of some catalysts, such as copper. It also has a strong tendency to absorb carbon dioxide and moisture. Neutralizations can be carried out in hydrazine just as in liquid ammonia. Acids in hydrazine are compounds containing the cation $NH_2NH_3^+$, while bases are compounds containing the anion NH_2NH^-. An example of a neutralization is therefore:

$$NH_2NH_3^+ + Cl^- + NH_2NH^- + Na^+ \longrightarrow 2\,NH_2NH_2 + Na^+ + Cl^-.$$

Solvolyses are also known in this solvent, e.g.

$$RC(O)OR + NH_2NH_2 \longrightarrow RC(O)NHNH_2 + ROH$$

$$FSO_2OK + 2NH_2NH_2 \longrightarrow H_2NNHSO_2OK + [NH_2NH_3]F.$$

2.9. Bibliography

H.H. Sisler: Chemistry in Non-Aqueous Solvents, Reinhold, New York 1961
W.L. Jolly u. C.J. Hallada: Liquid Ammonia, in T.C. Waddington: Non-Aqueous Solvent Systems. Academic Press, London, New York 1965
L.F. Audrieth u. J. Kleinberg: Non-Aqueous Solvents. Wiley, New York 1953, Chapter 3-6
G. Jander: Die Chemie in wasserähnlichen Lösungsmitteln. Springer, Berlin, Göttingen, Heidelberg 1949, Chapter 3
W.L. Jolly: Metal-Ammonia Solutions, in F.A. Cotton: Progress in Inorganic Chemistry. Interscience, New York 1959, Vol.1
J. Jander: Anorganische und allgemeine Chemie in flüssigem Ammoniak, in G. Jander, H. Spandau u. C.C. Addison: Chemie in nichtwässerigen ionisierenden Lösungsmitteln. Vieweg, Braunschweig 1966,Vol.I,1
H. Smith: Organic Reactions in Liquid Ammonia, in G. Jander, H. Spandau u. C.C. Addison: Chemie in nichtwässerigen ionisierenden Lösungsmitteln. Vieweg, Braunschweig 1963, Vol. I, 2
J. Jander, L. Bayersdorfer, K. Höhne, Z. anorg. allg. Chem. **357**, 215(1968)
H. Hartl, H. Bärnighausen, J. Jander, Z. anorg. allg. Chem. **357**, 225(1968)
U. Schindewolf, R. Vogelsgesang u. K.W. Böddeker: Angew. Chem. **79**, 1064 (1967)

3. Liquid Hydrogen Fluoride and the Higher Hydrogen Halides

3.1. Liquid Hydrogen Fluoride

The main difference between liquid ammonia and water is that the former exhibits a greater basicity. In this respect, indeed, it occupies a unique position among the common ionizing solvents. On the other hand, there are several solvents with a greater acidity. These include liquid hydrogen fluoride, the other liquid hydrogen halides, concentrated sulphuric acid, and even glacial acetic acid. The solvent theory of acids and bases, which we have seen applied to liquid ammonia, is equally applicable to these protonic solvents.

Despite its strong solvent power for inorganic and organic compounds, liquid hydrogen fluoride has only fairly recently been examined in any depth as an ionizing solvent. This is partly because of the great interest that fluorine chemistry in general has gained in connection with the exploitation of nuclear power (for example, uranium hexafluoride is used for the separation of uranium isotopes; liquid hydrogen fluoride is now used, not only as a solvent system, but also for the preparation of elementary fluorine and for the synthesis of inorganic and organic fluorine compounds). A second factor is that the extensive study of this solvent has only become possible with the use of fluorinated organic compounds for the production of apparatus. Substances such as glass and quartz are attacked by hydrogen fluoride, and until the introduction of fluorinated plastics such as polytetrafluoroethylene and polychlorotrifluoroethylene, which are resistant to hydrogen fluoride, it was necessary to use platinum or gold apparatus.

3.1.1. Physico-Chemical Properties of Liquid Hydrogen Fluoride

The physico-chemical properties that are important to the solvent characteristics of liquid hydrogen fluoride are shown in Table 7.

The small electrical conductivity of liquid hydrogen fluoride is explained by self-ionization as follows:

$$2\,HF \rightleftharpoons H_2F^+ + F^-$$

Though in view of the association of the solvent and the high mobility of the fluoride ion in liquid hydrogen fluoride (cf. Section 3.1.2), a chain conduction mechanism involving associated anions such as HF_2^-, $H_2F_3^-$ [$= (HF)_2F^-$], or $H_4F_5^-$ [$=(HF)_4F^-$] is more probable.

Table 7. Physico-chemical properties of liquid hydrogen fluoride

Melting point (°C)	−89.37
Boiling point (°C)	19.51
Heat of fusion (cal/g)	46.93 ± 0.04 (at the melting point)
Heat of vaporization	89.45 ± 0.2cal/g = 1.789 kcal/mole
	(at the boiling point)
Density (g/ml)	1.002 (at 0 °C)
Viscosity (c Poise)	0.256 (at 0 °C)
Dielectric constant	175 (at −73 °C)
	134 (at −42 °C)
	111 (at −27 °C)
	84 (at 0 °C)
Electrical conductivity	1.4×10^{-5} (at −15 °C)
($\Omega^{-1}cm^{-1}$)	$\sim 1 \times 10^{-6}$ (at 0 °C)

The high boiling point and the wide liquid range of hydrogen fluoride suggest that this solvent, like water and liquid ammonia, is associated via hydrogen bonds. The relatively low viscosity, on the other hand, shows the absence of three-dimensional crosslinking such as that resulting from association in water or concentrated sulphuric acid. X-ray studies and the infrared spectrum of solid hydrogen fluoride point to the presence of zig-zag chains that interact with one another to some extent. The hydrogen bonds for which a bond energy of 6.7 kcal/mole has been reported, are retained in liquid hydrogen fluoride. In addition to chain polymers, the liquid probably also contains cyclic associates, e.g. containing six hydrogen fluoride molecules (cf. Fig. 5). It is concluded from the relatively low heat of vaporization that the degree of association does not change appreciably on transition from the liquid to the gaseous state. The average number of hydrogen fluoride molecules per associate is reported to be 4.7 at 0 °C and 363.8 mmHg. The degree of association naturally decreases with falling pressure and rising temperature, and there are only 1.118 molecules per associate at 38.0 °C and 407 mmHg. Gaseous hydrogen fluoride is probably one of the least ideal gases. The association in the gas phase can be demonstrated directly by IR spectroscopic studies, the stretching vibrations of the associates giving a broad absorption between 3100 and 3900 cm^{-1}. The nature and degree of association are strongly influenced by ionizing impurities. Thus the density and electrical conductivity of liquid hydrogen fluoride are increased e.g. by the addition of water, and reach a maximum at a water content of about 12%. The density of such solutions increases suddenly on freezing. These changes in density and conductivity can be explained partly by a transition from cyclic to chain associates, and partly by ionization (cf. the next paragraph and Section 3.1.3).

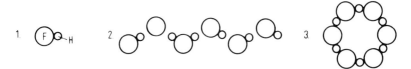

Fig. 5. Possible structures in solid and liquid hydrogen fluoride (from H.H.Hyman and J.J. Katz: Liquid Hydrogen Fluoride, in T.C. Waddington: Non-Aqueous Solvent Systems.Academic Press, London, New York 1965, p. 61, Fig. 8).

A few words on the acidity of hydrogen fluoride are now necessary. Liquid anhydrous hydrogen fluoride is a solvent with strong acidity, which protonates many inorganic and organic compounds containing a wide range of functional groups:

$$R + HF \rightleftharpoons RH^+ + F^-.$$

In dilute aqueous solution, on the other hand, hydrogen fluoride is a very weak acid. The explanation of this surprising situation is that though the stabilization of the anions iodide to fluoride by solvation increases with the electronegativity of the halogen, the stabilization of the hydrogen halide molecule increases with the square of the difference in electronegativity between the halogen and hydrogen. The acid strength of aqueous hydrogen halides is therefore greatest in hydrogen iodide and least in hydrogen fluoride. Though this qualitatively explains the weakness of dilute aqueous hydrogen fluoride solutions, the very low value of the equilibrium constant

$$K = \frac{[H_3O^+][F^-]}{[HF]} = 3 \times 10^{-4} \text{ to } 7 \times 10^{-4}.$$

is still surprising.

The increase in the strength of aqueous hydrogen fluoride solutions with increasing hydrogen fluoride concentration can also be explained as follows. Neither liquid hydrogen fluoride nor liquid water contains isolated molecules, but only higher aggregates. A proton transfer from an isolated hydrogen fluoride molecule to associated water in dilute aqueous hydrofluoric acid is a very different matter from the transfer of a proton from associated hydrogen fluoride to an isolated water molecule in concentrated liquid hydrogen fluoride. The first transfer gives only a solvated fluoride ion, while in the second case the fluoride ion remains a part of the relatively

stable zig-zag chain. Since the chains readily give up protons, concentrated hydrogen
fluoride solutions are more strongly acidic than dilute solutions.

A word on acidity scales for solvents might not be out of place here. The best
known acidity scale is that proposed by Hammett, which is particularly suitable for
solvents having high acidity and dielectric constants of between about 50 (anhydrous
formic acid) and 110 (anhydrous sulphuric acid). This scale is constructed with the
indicators (symbol In), which, as bases, are converted into the conjugate acids (symbol
InH^+) by the solvent. The ratio c_{InH^+}/c_{In} can generally be determined spectro-
photometrically or colorimetrically; one need only use an indicator whose unionized
form is present in measurable quantities in the medium in question. Thus a very weak
indicator base (e.g. substituted nitroanilines or azobenzenes) must be used. The dissociation constant K_{InH^+} of the indicator is given by

$$K_{InH^+} = \frac{a_{H^+} \cdot a_{In}}{a_{InH^+}} = \frac{c_{In}}{c_{InH^+}} \cdot \frac{\gamma_{In}}{\gamma_{InH^+}} \cdot a_{H^+}$$

where γ denotes activity coefficients. A simple rearrangement gives

$$a_{H^+} = K_{InH^+} \cdot \frac{c_{InH^+}}{c_{In}} \cdot \frac{\gamma_{InH^+}}{\gamma_{In}} .$$

Since the value $\gamma_{InH^+}/\gamma_{In}$ in a solution containing measurable quantities of both
forms of the indicator is unknown, the hydrogen ion activity cannot be calculated.
To a very good approximation, however the ratio $\gamma_{InH^+}/\gamma_{In}$ is independent of the
indicator used, but is characteristic of the medium. The expression $a_{H^+} (\gamma_{In}/\gamma_{InH^+})$
can therefore be taken as a quantity that is characteristic of the solvent. This quantity
is denoted by h_o, and unlike a_{H^+}, it can be measured:

$$h_o = K_{InH^+} \cdot \frac{c_{InH^+}}{c_{In}} .$$

h_o is thus a concentration-based measure of the ability of the solvent to convert
basic neutral molecules into their conjugate acids. By analogy with the pH, the
quantity H_o is defined as the negative logarithm of h_o:

$$H_o = - \log h_o .$$

This leads to the relationship:

$$H_o = (pK_{InH^+}) - \log \frac{c_{InH^+}}{c_{In}} .$$

Some typical H_o values are $+7.0$ for water, -2.0 for 6.9M hydrochloric acid, -10.03 for 96% sulphuric acid, and -12.08 for 100% sulphuric acid. Values around -10.5 are reported for liquid hydrogen fluoride. Thus anhydrous sulphuric acid and anhydrous hydrogen fluoride are roughly equivalent on the Hammett acidity scale. Both show a great acidity. Suitable indicators in these media include the very weak bases picramide ($=2.4.6$-trinitroaniline, $pK_{InH^+} = -9.29$), anthraquinone ($pK_{InH^+} = -8.2$), and p-nitroazobenzene ($pK_{InH^+} = -3.3$); thus very weak bases have very strongly negative pK_{InH^+} values.

3.1.2. Solubility of Inorganic and Organic Substances in Liquid Hydrogen Fluoride

Liquid hydrogen fluoride is finding increasing use as a solvent because of its low molecular weight, its high dielectric constant and its high acidity. It is also of interest as a solvent for very strong oxidizing agents, since it cannot be oxidized to elementary fluorine. However, reductions with liberation of hydrogen are possible. A number of qualitative solubility data are given in Table 8. Table 9 gives quantitative solubility data for some metal fluorides.

Finally, the solubilities of some classes of organic compounds are given in table 10. The dissolution of most substances is accompanied by ionization, as is shown by electrical conductivity measurements. The interpretation of this ionization in terms of the solvent theory of acids and bases will be discussed in Section 3.1.3. For the present we shall be concerned with simple facts.

1. The simplest situation is found with the metal fluorides, which ionize in accordance with the scheme:

$$HF + MF_x \rightleftharpoons MF_{x-1}^+ + HF_2^-.$$

The HF_2^- ion, which also occurs in acid salts such as HF. KF, is a symmetrical ion, i.e. the proton is situated centrally between the two fluoride ions (hydrogen bonding). Conductivity measurements on solutions of potassium and sodium fluorides gave a molar conductivity of about $400\Omega^{-1}cm^2mole^{-1}$ at $0\,^{\circ}C$. About 70% of the current transport appears to be effected by HF_2^- ions. These ions therefore move by a chain mechanism similar to that of the proton or of the hydroxide ion in water, However, $H_2F_3^-$ and $H_4F_5^-$ ions are probably present as well as HF_2^- (cf. Section 3.1.1.).

2. Another dissolution mechanism, which applies mainly to organic substances, but is also observed with inorganic proton acceptors, involves protonation in accordance with

$$R + 2HF \longrightarrow RH^+ + HF_2^-.$$

Table 8. Solubilities of some groups of substances in liquid hydrogen fluoride

Solute	Solubility
Alkali metal, silver, thallium fluorides	Abundantly soluble
Alkaline earth metal fluorides	Fairly abundantly soluble
Heavy metal fluorides (and zinc and aluminium fluorides)	Insoluble
Hydrogen halides	Insoluble
Alkali and alkaline earth metal halides	Formation of the corresponding fluorides, liberation of hydrogen halide
Heavy metal halides	Insoluble
Metal oxides, hydroxides (e.g. aluminium oxide, copper (I) oxide)	Formation of the corresponding fluorides, simultaneous formation of water
Heavy metal oxides	Insoluble
Alkali metal and silver nitrates	Abundantly soluble
Heavy metal nitrates	Insoluble
Alkali and alkaline earth metal sulphates	Soluble
Heavy metal sulphates	Insoluble
Non-metal fluorides: boron trifluoride, tin (IV) fluoride, phosphorus, arsenic and antimony (V) fluorides, chlorine and bromine trifluorides	Abundantly soluble

Table 9. Solubilities of some metal fluorides in liquid hydrogen fluoride

Fluoride	Solubility (g/100 g HF)	Temp. ($^{\circ}$C)	Fluoride	Solubility (g/100 g HF)	Temp. ($^{\circ}$C)
LiF	10.3	12	AgF_2	0.048	12
NaF	30.1	11	CaF_2	0.817	12
KF	36.5	8	PbF_2	2.62	12
RbF	110.0	20	SbF_3	0.536	12
CsF	199.0	10	SbF_5	–	25
NH_4F	32.6	17	NbF_5	6.8	25
AgF	83.2	19	TaF_5	15.2	25

Table 10. Solubilities of organic substances in liquid hydrogen fluoride

Solute	Solubility
Pure hydrocarbons (open-chain and cyclic)	Slightly soluble
Polyenes	Practically insoluble
Alkyl fluorides	Abundantly soluble
Alkyl and aryl halides	Insoluble
Aliphatic alcohols	Abundantly soluble
Aromatic alcohols	Abundantly soluble
Aliphatic and aromatic aldehydes and ketones	Abundantly soluble
Aliphatic ethers	Abundantly soluble
Aromatics ethers	Slightly soluble
Aliphatics and aromatic acids	Mostly abundantly soluble
Acyl fluorides	Abundantly soluble
Acyl halides	Abundantly soluble
Acid anhydrides	Abundantly soluble
Amines	Abundantly soluble
Carbohydrates and cellulose	Abundantly soluble

This protonation is a very common process, owing to the acidity of liquid hydrogen fluoride. Weak proton acceptors, the protophilicity of which can be still further reduced by substituents such as nitro or trifluoromethyl groups, are protonated only to a very small extent, and the electrical conductivity of their solutions is only slightly higher than that of the pure solvent. The proton transfer results from the formation of hydrogen bonds, which are therefore important to our understanding of solutions in liquid hydrogen fluoride. It is difficult to explain the ready solubility of extremely weak proton acceptors such as sulphur dioxide or trifluoroacetic acid; these are miscible with liquid hydrogen fluoride in all proportions, and can be used as diluents for this solvent, since they do not appreciably influence its acidity. Some organic substances also dissolve in liquid hydrogen fluoride without appreciable ionization. These include the chain and cyclic hydrocarbons.

3. A third, less common ionization mechanism that can make a substance soluble is the addition of fluoride ions in accordance with

$$MF_x + 2\,HF \longrightarrow MF_{x+1}^- + H_2F^+.$$

This mechanism applies mainly to the fluorides of the elements of the higher main groups in the periodic system and to some transition metal fluorides. Where steric factors make the addition of fluoride ions impossible, as in some hexafluorides, the compound is either insoluble or does not ionize. Thus uranium hexafluoride is only very slightly soluble in liquid hydrogen fluoride. An exception to this general rule is xenon hexafluoride, whose ready solubility, fast fluorine exchange, and high degress of ionization point to a symmetry lower than that of the octahedron.

3.1.3. Acids, Bases, and Salts

3.1.3.1. Acids and Bases

According to the solvent theory of acids and bases, acids are substances that increase the concentration of solvent cations, whereas bases increase the concentration of the solvent anions. From the self-dissociation of hydrogen fluoride

$$3\ HF\ \leftrightharpoons\ H_2F^+ + HF_2^-$$

all substances that increase the concentration of hydrogen ions (or fluoronium ions) will be acids, and all substances that increase the concentration of fluoride ions (or HF_2^- ions) will be bases.

Owing to the strong acidity of liquid hydrogen fluoride, the only known substances that increase the concentration of fluoronium ions are perchloric acid and fluorosulphuric acid; thus solvo acids are practically non-existent. Even such strong acids as nitric and sulphuric acids behave as ansolvo bases:

$$HONO_2 + 2\ HF\ \leftrightharpoons\ (HO)_2NO^+ + HF_2^-$$

$$(HO)_2SO_2 + 3\ HF\ \longrightarrow\ HOSO_2F + H_3O^+ + HF_2^-.$$

The weaker acetic acid is naturally also an ansolvo base in liquid hydrogen fluoride:

$$CH_3COOH + 2\ HF\ \leftrightharpoons\ CH_3C(OH)_2^+ + HF_2^-.$$

Though solvo acids are practically unknown, it is still possible for substances to increase the concentration of fluoronium ions by capture of fluoride ions, and hence to act as ansolvo acids. It has recently been found that the fluorides of elements of main groups 4,5, and 6 in the periodic system behave in this way. Conditions for the capture of further fluoride ions are that bonding orbitals should be available on the central atom and that there should be a suitable geometrical configuration (coordination number) available. Thus neither carbon tetrafluoride nor sulphur hexafluoride reacts as an ansolvo acid. On the other hand, the fluorides of phosphorus, arsenic, and antimony, PF_5, AsF_5, and SbF_5 are fluoride ion acceptors. The boiling points and Trouton constants indicate that the most highly associated of these is antimony pentafluoride. Since the normal coordination number of antimony toward fluoride ions is six, antimony pentafluoride is a particularly strong ansolvo acid in liquid hydrogen fluoride. It reacts to give H_2F^+ ions

$$SbF_5 + 2\,HF \rightleftharpoons H_2F^+ + SbF_6^- \,,$$

as is shown by the marked increase in the electrical conductivity. Arsenic pentafluoride and phosphorus pentafluoride, on the other hand, are only weak acids in liquid hydrogen fluoride.

The same is true of boron trifluoride, tin and germanium tetrafluorides, and iodine pentafluoride:

$$BF_3 + 2\,HF \rightleftharpoons BF_4^- + H_2F^+$$

$$SnF_4 + 4\,HF \rightleftharpoons SnF_6^- + 2\,H_2F^+$$

$$IF_5 + 2\,HF \rightleftharpoons IF_6^- + H_2F^+.$$

Some fluorides of transition metals of the group 5 and 6 triads are also weak acids. Titanium and silicon tetrafluorides are insoluble, and do not increase the concentration of fluoronium ions. In multiphase systems with hydrocarbons however, titanium tetrafluoride increases the solubility of some hydrocarbons and acts as a catalyst for their rearrangements. This behaviour resembles that of tantalum and molybdenum pentafluorides in similar systems, and is attributed to the acidic character of the fluorides. The solubility and acidstrength of sparingly soluble and only slightly acidic fluorides is increased by proton acceptors (R, e.g. amines):

$$R + TaF_5 + HF \longrightarrow RH^+ + TaF_6^-$$

$$R + 2\,TiF_4 + HF \longrightarrow RH^+ + Ti_2F_9^-.$$

The small number of acids in liquid hydrogen fluoride, most of which are ansolvo acids, contrasts with the many solvo and ansolvo bases. According to the solvent theory, all soluble fluorides that increase the concentration of fluoride (or HF_2^-)

ions are solvo bases. The alkali metal fluorides are strong bases, as are the fluorides of elements of main group 7 of the periodic system, such as chlorine trifluoride and bromine trifluoride:

$$ClF_3 + HF \longrightarrow ClF_2^+ + HF_2^-$$

$$BrF_3 + HF \longrightarrow BrF_2^+ + HF_2^- \, .$$

Nitric, sulphuric, and acetic acids have already been mentioned as examples of ansolvo bases. Trifluoroacetic acid, which is a much stronger acid than acetic acid, is a weaker ansolvo base than the latter in liquid hydrogen fluoride. Similarly, trifluoroethanol is a weaker ansolvo base than ethanol, as is shown by the lower conductivity of its solutions. Other ansolvo bases are aldehydes, ketones, ethers, amines, and orgar compounds containing sulphur. Liquid hydrogen fluoride is a poor solvent for saturated hydrocarbons, whereas unsaturated hydrocarbons can act as proton acceptors. In many cases they can be recovered unchanged, but polymerizations and rearrangen are also possible, as e.g. in the case of butadiene. The solutions are often strongly coloured. Aromatic and substituted aromatic compounds dissolve readily in liquid hydrogen fluoride, and react as weak bases. For example, nitrobenzene, which is used as an indicator for the construction of the Hammett acidity scale (see Section 3.1.1), conducts electricity in liquid hydrogen fluoride.

A few observations concerning the ansolvo base water may be noted here. The addition of a little water to liquid hydrogen fluoride leads to an increase in electrical conductivity. The water is protonated:

$$H_2O + 2 HF \longrightarrow H_3O^+ + HF_2^- \, ,$$

but the solubility of the solution is less e.g. than that of solutions of the solvo bases potassium and sodium fluorides or of the ansolvo bases methyl alcohol and acetic acid at comparable concentrations. It must therefore be assumed that the water molecules are embedded in associates, which are held together by hydrogen bonds, and each of which has only one proton:

$$nH_2O + 2 HF \longrightarrow (H_2O)_n H^+ + HF_2^-$$

$$nH_2O + mHF \longrightarrow (HF)_{m-2}(H_2O)_n H^+ + HF_2^- \, .$$

Despite the difference in the acidities of liquid hydrogen fluoride and water, there is a certain similarity between the acid-base behaviour of the fluorides in the former an that of the oxides in the latter. The acid strength increases steadily from the basic alkali metal compounds to the compounds of the elements of groups 5 and 6:

Strong bases: $NaF + HF \longrightarrow Na^+ + HF_2^-$

$Na_2O + H_2O \longrightarrow 2\,Na^+ + 2\,OH^-$

Weak acids: $AlF_3 + HF + NaF \longrightarrow Na^+ + AlF_4^- + HF$

$Al_2O_3 + 3\,H_2O + 2\,NaOH \longrightarrow 2\,Na^+ + 2\,Al(OH)_4^-$

Strong acids: $SbF_5 + 2\,HF \longrightarrow H_2F^+ + SbF_6^-$

$SO_3 + 3\,H_2O \longrightarrow 2\,H_3O^+ + SO_4^{2-}$

$Cl_2O_7 + 3\,H_2O \longrightarrow 2\,H_3O^+ + 2\,ClO_4^-.$

Only a few compounds of the elements of group 7 deviate noticeable from this analogy. The halogen fluorides ClF_3 and BrF_3 are solvo bases in liquid hydrogen fluoride, while the corresponding halogen oxides are ansolvo acids in water.

3.1.3.2. Salt formation by Neutralization and by Dissolution of Metals in Acids

Since acids and bases exist in liquid hydrogen fluoride, the process of neutralization is possible. This consists in the combination of fluoronium and fluoride ions:

$$H_2F^+ + HF_2^- \longrightarrow 3\,HF.$$

An example of a neutralization is the formation of fluorobromonium hexafluoroantimonate:

$$BrF_2^+ + HF_2^- + H_2F^+ + SbF_6^- \longrightarrow BrF_2SbF_6 + 3\,HF.$$

Hexafluoroantimonic acid reacts similarly with the base water, and this reaction can be followed conductimetrically:

$$H_3O^+ + HF_2^- + H_2F^+ + SbF_6^- \longrightarrow H_3OSbF_6 + 3\,HF.$$

Silver fluoride gives silver hexafluoroantimonate:

$$AgF + H_2F^+ + SbF_6^- \longrightarrow AgSbF_6 + 2\,HF.$$

The alkali metal salts of this acid can also be prepared, and isolated in the pure state by distillation of the solvent from the reaction mixture. Like sodium hexahydroxo-antimonate in water, sodium hexafluoroantimonate is sparingly soluble in liquid hydrogen fluoride, and precipitates out on addition even of small quantities of sodium fluoride to a solution of fluoronium hexafluoroantimonate.

Many salts can also be prepared by the dissolution of metals by acids in liquid hydro$_?$
fluoride.

The dissolution of increasing basic metal fluorides ($CoF_3 < MnF_3 < CrF_3 <$
$HgF_2 < AgF_2 < CuF_2 < NiF_2 < CoF_2 < CaF_2 < AgF < NaF$) and of metals of increas
electropositivity ($Hg < Ag < Cu < Pb < Sn < Nb < Cr < Zn < Mn < Mg < Ca$) by
acids in liquid hydrogen fluoride is also a measure of the strength of the dissolved
acid. Thus dissolution of heavy metal fluorides (with the exception of silver (I)
fluoride) indicates that the quantity of fluoride ions given by the solubility product
is decreased, and hence that the solution acting on the fluoride is acidic. The solubility of cobalt (III) fluoride or of zinc in solutions of antimony (V) fluoride thus
demonstrates the acidic character of this pentafluoride. The dissolution of metals in
acids with evolution of hydrogen naturally also depends on whether the salt film for
on the surface of the metal dissolves and whether there is a hydrogen overvoltage on
the metal. The hydrogen overvoltage is generally lower in liquid hydrogen fluoride
than in water. For example, thallium does not dissolve in water because of the overvoltage of 0.41 V, but dissolves rapidly in liquid hydrogen fluoride with evolution of
hydrogen. The overvoltage at mercury electrodes is also much lower in liquid hydrog
fluoride (0.4V) than in water (about 2V). The solubility of noble metals in acids
depends further on whether dissolution is accompanied by liberation of hydrogen or
by "oxidation" (=fluorination in liquid hydrogen fluoride), as in an electromotive
series. To dissolve silver and mercury, for example, the ansolvo acid must have an
"oxidizing action" (i.e. a fluorinating action). If no such action is to be expected, the
ansolvo acid should not react with the noble metals. Thus fluoroboric acid dissolves
magnesium, but does not attack the nobler metals. The ansolvo acids VF_5, IF_5, TeF_6,
MoF_6, WF_6, and ReF_6, on the other hand, react with silver and mercury because of
their "oxidizing action", e.g.

$$6\,IF_6^- + 6\,Ag + 6\,H_2F^+ \longrightarrow 5\,AgIF_6 + AgI + 12\,HF$$

$$2\,TeF_7^- + 3\,Sn + 8\,HF \longrightarrow 3\,SnF_6^{--} + 2\,Te + 4\,H_2F^+.$$

The failure of these acids to react with copper and the more electropositive metals is
due to the fact that they are weak acids, and cannot dissolve the fluoride layer form
on such metals by their fluorinating action:

$$IF_6^- + 3\,Mn + H_2F^+ \longrightarrow 3\,MnF_2 + HI + HF.$$

In the case of silver, the silver fluoride formed is a strong and relatively soluble base,
and a precipitate is formed only when the concentration of silver hexafluoroiodate
reaches the value given by the solubility product. Ansolvo acids with low acid

strengths and weak "oxidizing power" normally do not react with metals. Examples of such ansolvo acids are titanium (IV) fluoride, germanium (IV) fluoride, and selenium (IV) fluoride.

There are many examples of the formation of salts by dissolution of basic fluorides or metals in solutions of ansolvo acids in liquid hydrogen fluoride. Some of these will now be discussed.

Neodymium trifluoride dissolves in a solution of antimony pentafluoride to form the corresponding salt:

$$NdF_3 + 3 H_2F^+ + 3 SbF_6^- \longrightarrow Nd(SbF_6)_3 + 6 HF.$$

Neodymium (III) oxide may also be used in this case. The weak base neodymium fluoride can be reprecipitated from the resulting salt solution by the stronger base sodium fluoride:

$$3 NaF + Nd(SbF_6)_3 \longrightarrow NdF_3 \downarrow + 3 NaSbF_6.$$

On the basis of salt-forming reactions of this type, hexafluoroantimonic acid is found to be the strongest acid in the liquid hydrogen fluoride system. However, since the cobalt salt of this acid is as highly solvolysed at high dilutions as iron or aluminium acetate in water, the acid strength of antimony pentafluoride in relation to liquid hydrogen fluoride is no greater than that of acetic in relation to water.

As was mentioned earlier, sulphur hexafluoride is not an ansolvo acid, since steric factors prevent the addition of further fluoride ions. The products obtained on fluorination of selenium and tellurium with chlorine trifluoride in liquid hydrogen fluoride, on the other hand, exhibit weakly acidic properties. Tellurium in the tetravalent state forms the ion TeF_6^{2-}, while salts such as $AgTeF_7$ can be prepared from the hexavalent form. The acid involved is so weak that it does not dissolve weak bases such as the fluorides CoF_3, CuF_2, and MnF_3 and does not attack metals such as magnesium, manganese, and chromium.

It can therefore be seen that, contrary to ealier views, acids exist in liquid hydrogen fluoride, and acid-base phenomena fully comparable to those in water and in liquid ammonia can be observed.

3.1.4. Amphoterism, Solvolyses, Solvates (Hydrogen Fluoride Adducts)

Though chemistry in liquid hydrogen fluoride is poorer in acids and richer in bases than that in water or liquid ammonia, comparable acid-base reactions nevertheless take place. Similarly, amphoterism, solvolysis, and solvate formation in these solvent systems are also comparable to some extent, i.e. phenomena of this type also occur in liquid hydrogen fluoride.

Thus the base aluminium fluoride, like aluminium hydroxide, exhibits amphoteric character. It dissolves in excess sodium fluoride in hydrogen fluoride to give cryolite Na_3AlF_6. On addition of the acid boron trifluoride, white aluminium fluoride precipitates out again. Similar behaviour is observed with chromium (III) fluoride, and antimony (III) fluoride (soluble in the presence of excess sodium fluoride). Zinc, iron (III), and cobalt (III) fluorides, on the other hand, are not amphoteric.

Solvolysis is a particularly common type of reaction in liquid hydrogen fluoride. One of the few salts that are not solvolysed, i.e. that behave as salts of strong acids, is potassium perchlorate, which (as in water) dissolves without reacting with the solvent

$$KClO_4 \rightleftharpoons K^+ + ClO_4^-.$$

Sulphates, on the other hand, are converted into fluorsulphates. The alkali metal halides are completely solvolysed, since the resulting gases hydrogen chloride, bromide and iodide are practically insoluble in liquid hydrogen fluoride, and are in any case extremely weak acids:

$$KCl + 2\,HF \longrightarrow K^+ + HF_2^- + HCl \uparrow$$

$$RbBr + 2\,HF \longrightarrow Rb^+ + HF_2^- + HBr \uparrow.$$

(Nevertheless, hydrogen chloride is so soluble near the melting point of hydrogen fluoride ($-89\,°C$) that silver chloride can be precipitated.) Similarly, salts such as potassium acetate and potassium nitrate are solvolysed with formation of the free acids; however, these can react further as ansolvo bases, so that the electrical conductivity observed corresponds to four ions per molecule of salt:

$$CH_3COOK + 2\,HF \longrightarrow CH_3COOH + K^+ + HF_2^-$$

$$CH_3COOH + 2\,HF \longrightarrow CH_3C(OH)_2^+ + HF_2^-$$

$$KNO_3 + 3\,HF \longrightarrow K^+ + 2\,HF_2^- + HNO_3 \cdot H^+\,(NO_2^+ \cdot H_2O).$$

Thus the dissolution of salts in liquid hydrogen fluoride nearly always results in the formation of HF_2^- ions, i.e. in solvolytic base formation.

Potassium nitrite dissolves in liquid hydrogen fluoride at high temperatures, mainly with liberation of oxides of nitrogen. On the other hand, if the nitrite is added carefully to strongly cooled hydrogen fluoride, it forms solutions from which hardly any nitrogen oxide escapes and whose content of trivalent nitrogen remains unchanged for long periods. It therefore appears that a process similar to that found on dissolution of nitrites in concentrated sulphuric acid or perchloric acid and leading to the formation of nitrosylsulphuric acid or nitrosyl perchlorate occurs:

$$KNO_2 + 3\,HF \longrightarrow K^+ + 2\,HF_2^- + HNO_2 \cdot H^+\,(NO^+ \cdot H_2O).$$

This reaction also corresponds to that observed on dissolution of potassium nitrate in liquid hydrogen fluoride (see above). As in sulphuric acid or perchloric acid, the UV spectrum of nitrous acid can be seen to change into that of the nitric oxide cation $NO^+ \cdot H_2O$ with decreasing water content. Nitrous acid thus reacts as an ansolvo base in all three acids:

$$NOOH + H_2SO_4 \longrightarrow NO^+H_2O + HSO_4^-$$

$$NOOH + HClO_4 \longrightarrow NO^+ \cdot H_2O + ClO_4^-$$

$$NOOH + 2\,HF \longrightarrow NO^+ \cdot H_2O + HF_2^-.$$

The electrical conductivity of a solution of nitrous acid is twice (three times in the case of potassium nitrite) that expected for a 1,1 electrolyte. The cation $NO^+ \cdot H_2O$ is thus further protonated in liquid hydrogen fluoride.

$$NO^+ \cdot H_2O + 2\,HF \longrightarrow NO^+ + H_3O^+ + HF_2^-.$$

The existence of the nitrosyl cation in solutions of nitrous acid in hydrogen fluoride is also shown by the formation of salts containing this cation, e.g.:

$$NO^+ + F^- + H_2F^+ + SbF_6^- \longrightarrow 2\,HF + NOSbF_6.$$

Moreover, the weak base nitrosyl fluoride can be driven off by the addition of a strong base such as potassium fluoride:

$$NO^+ + F^- + K^+ + F^- \longrightarrow NOF{\uparrow} + K^+ + F^-.$$

If nitric oxide is passed into such solutions, they assume a deep indigo colour ("violet hydrogen fluoride") and exhibit the characteristic thermochromism of the nitric oxide-nitrosyl ion on cooling:

$$NO^+ + NO \longrightarrow N_2O_2^+.$$

These solutions are similar to those formed by introduction of nitric oxide into nitrosylsulphuric acid or by partial reduction of nitrosylsulphuric acid with sulphur dioxide or methanol. Like the nitrites, dinitrogen trioxide is solvolysed in liquid hydrogen fluoride with initial formation of nitrosyl fluoride and nitrous acid:

$$(N_2O_3 \rightleftharpoons NO^+NO_2^-) + HF \longrightarrow NOF + HNO_2.$$

The overall process that takes place is as follows:

$$N_2O_3 + 3\,HF \longrightarrow 2\,NO^+ + H_3O^+ + 3\,F^-.$$

Distillation of solutions of metal nitrites and of dinitrogen trioxide in liquid hydrogen fluoride at 68 °C gives a product having the composition $N_2O_3 \cdot 12.8\,HF$ or NOF $\cdot H_2O$. This is not an azeotropic mixture, since on removal of the water with sulphur tetrafluoride:

$$SF_4 + H_2O \longrightarrow SOF_2 + 2\,HF$$

followed by elimination of the resulting thionyl fluoride, a product having the composition NOF $\cdot$ 3 HF can be isolated by distillation at 94 °C. Like the product that distils over at 68 °C, nitrosyl fluoride-3-hydrogen fluoride exhibits an electrical conductivity of the same order of magnitude as that of salt melts. Pure nitrosyl fluoride-3-hydrogen fluoride is a colourless hygroscopic liquid, which can be stored in Teflon, paraffin, or pure nickel containers and is a good fluorinating agent.
The above discussion readily explains the formation of nitrosyl fluoride and nitric acid on solvolysis of nitrogen dioxide in liquid hydrogen fluoride:

$$(2\,NO_2 \rightleftharpoons N_2O_4 \rightleftharpoons NO^+NO_3^-) + HF \longrightarrow NOF + HNO_3.$$

As in the cases of nitrites and dinitrogen trioxide, further protonation steps follow. Compounds that already contain the nitrosyl cation, such as nitrosyl chloride, are also solvolysed in liquid hydrogen fluoride:

$$NOCl + HF \longrightarrow NOF + HCl.$$

This reaction is favoured by the complex formation between nitrosyl fluoride and hydrogen fluoride molecules.
In the discussion of solvolysis in liquid ammonia, it was mentioned that solvolysis is preceded by adduct formation, and that many adducts can be isolated. Adducts can also isolated in the hydrogen fluoride system, though these are far fewer than the water or ammonia adducts. Moreover, all the known hydrogen fluoride adducts are adducts of fluorides. Thus there are solvates having the general formula MF $\cdot$ HF (M = alkali metal or thallium) or MF $\cdot$ 2 HF (M = potassium, rubidium, or thallium), the existence of which confirms the presence of the ions HF_2^- and $H_2F_3^-$ predicted from conductivity measurements (cf. Section 3.1.1 and 3.1.2). Trisolvates MF $\cdot$3 HF

(M = potassium, rubidium, or silver), which include the nitrosyl fluoride-3-hydrogen fluoride discussed in the last paragraph, are also known. Mixed water-hydrogen fluoride solvates (e.g. $CuF_2 \cdot 5\,HF \cdot 6\,H_2O$, $CaF_2 \cdot 2\,HF \cdot 6\,H_2O$, and $2\,AlF_3 \cdot HF \cdot 5\,H_2O$) have also been observed.

3.1.5. Electrofluorination of Organic Compounds in Liquid Hydrogen Fluoride

Liquid hydrogen fluoride is of great interest at present as a medium for the preparation of fluorinated organic compounds by electrofluorination. Copper and iron are used as container materials; the cathode consists of copper, iron, or nickel and the anode of nickel. The electrolysis is carried out at a current density of about $0.04\,A/cm^2$ an a voltage of $5 - 7$ V. Electrofluorination is normally applicable only to compounds having low volatilities, which can be dissolved, suspended, or emulsified in liquid hydrogen fluoride. Gaseous substances can also be used if they can be made to rise between the electrodes, which are situated $2 - 5$ mm apart. The electrical conductivity of the electrolyte is increased by the addition of 5% potassium or sodium fluoride. The detailed mechanism of the electrolysis is not yet known, particularly since hydrogen fluoride is not electrolytically decomposed under the conditions used. Several products are formed in many cases. For example, the electrofluorination of octane yields perfluorooctane C_8F_{18}, perfluoroethane C_2F_6, and perfluoromethane CF_4. The electrolysis of propane in liquid hydrogen fluoride leads mainly to perfluoropropane C_3F_8, together with $CF_3-CHF-CF_3$ and $CF_3-CF_2-CHF_2$, Ethane gives perfluoroethane C_2F_6 together with C_2F_5H and $C_2F_4H_2$. Methane gives all four possible fluorination products. The electrolysis of hydrogen sulphide in hydrogen fluoride containing water yields sulphuryl fluoride, thionyl fluoride, and sulphur tetrafluoride, but gives sulphur hexafluoride in the absence of water. The results of other fluorinations are given in Table 11.

Table 11. Products of the electrofluorination of some organic compounds.

Group of compounds	Starting material	Fluorination products
Alcohols	C_2H_5OH	C_2F_6
Ethers	$C_2H_5-O-CH_3$	$C_2F_5-O-CF_3$, C_2F_6, CF_4
Carboxylic acids	CH_3COOH	CF_3COF, CF_4, CF_3H, CF_2H_2, CFH_3, CO_2, OF_2
Organic compounds containing nitrogen	CH_3NH_2	CF_4, NF_3
	$(CH_3)(C_2H_5)(C_3H_7)N$	$(CF_3)(C_2F_5)(C_3F_7)N$, CF_4, C_2F_6, C_3F_8

The importance of electrofluorination, and hence of the use of liquid hydrogen fluoride as a solvent, can be seen from the fact that trifluoroacetic acid CF_3COOH, which is obtained by hydrolysis of trifluoroacetyl fluoride CF_3COF, is used as a catalyst for esterifications of alcohols and phenols with aliphatic and aromatic carbo-xylic acids, as a condensing agent in the preparation of ketones and sulphones, as a starting material for the preparation of trifluoroiodomethane CF_3I (which is in turn a starting material for the preparation of many compounds in carbon-fluorine che-mistry), and as an important intermediate for the synthesis of organic fluorine compounds.

3.1.6. Closing Remarks on the Hydrogen Fluoride System

The discussion of reactions in liquid hydrogen fluoride shows the similarity between this solvent and water, but also demonstrates the considerable difference between them, which is mainly due to the strong acidity of hydrogen fluoride.

Like water, liquid hydrogen fluoride dissolves many inorganic and organic substan-ces. The pure solvent, like water, has a slight electrical conductivity. Most substances are dissociated in liquid hydrogen fluoride, just as in water, and adduct formation is observed in both solvents. However, since the fluoride ion is the most stable ion in liquid hydrogen fluoride, the only adducts known are of fluorides. Amphoteric beha-viour is also observed in some cases.

Unlike water, liquid hydrogen fluoride exhibits few types of reactions. The most common types are solvolysis and protonation; consequently, the fluoride or HF_2^- ion is by far the most commonly occurring anion, and bases (ansolvo and solvo bases) occur particularly frequently. The range of available cations is therefore greater than in water, but the number of precipitation reactions is limited by the lack of variety in the anions. Some further anions are formed by the few known ansolvo acids, whereas solvo acids are practically unknown because of the acidity of liquid hydrogen fluoride.

3.2. The Higher Hydrogen Halides

3.2.1. Physico-Chemical Properties of the Higher Hydrogen Halides

Liquid hydrogen chloride, bromide and iodide are of some interest as ionizing solvent Their properties are similar in some respects to those of liquid hydrogen fluoride, but differences also exist, as can be seen from Table 12.

Table 12. Physico-chemical properties of water and of liquid hydrogen halides

	H_2O	HF	HCl	HBr	HI
Molecular weight	18.0	20.0	36.5	80.9	127.9
Melting point (°C)	0	− 83.0	−114.6	− 88.5	− 50.9
Boiling point (°C)	100	19.5	−84.1	− 67.0	− 35.0
Liquid range (°C)	100	102.5	30.5	21.5	15.9
Heat of fusion (cal/mole)	1440	1094	476	600	686
Heat of vaporization (cal/mole)	9720	7230	3860	4210	4724
Trouton constant	26.0	24.7	20.4	20.4	19.8
Dielectric constant	84.2 (0 °C)	175 (−73 °C)	9.28 (−95 °C)	7.0 (−85 °C)	3.39 (−50 °C)
Viscosity (cPoise)	1.0 (22 °C)	0.24 (0 °C)	0.51 (−95 °C)	0.83 (−67 °C)	1.35 (−35,4 °C)

The higher hydrogen halides have much narrower liquid ranges than hydrogen fluoride, so that investigations in these solvents are more difficult. The values of the Trouton constant are normal, indicating only slight association; the hydrogen bonds must therefore be much weaker than those in liquid hydrogen fluoride. These solvents also have relatively low dielectric constants, with the result that they only dissolve substances having low lattice energies. Thus while the tetraalkylammonium halides are soluble, ammonium chloride is not, since it has a higher lattice energy. Owing to their conductivity (the specific conductivites are one tenth to one hundredth of that of water), the higher hydrogen halides must be assumed to undergo self-ionization:

$$3 HX \rightleftharpoons H_2X^+ + HX_2^-.$$

Whereas the ions H_2F^+ and HF_2^- are known in the case of hydrogen fluoride, the existence of corresponding ions in the higher hydrogen halides is rather uncertain. The cation H_2Cl^+ is assumed to exist in the explosive salt $HCl \cdot HClO_4$. The existence of H_2Br^+ and H_2I^+ in solution has not yet been demonstrated. The anion HCl_2^- occurs in several compounds, including the salts $(CH_3)_4N^+HCl_2^-$ and $Cs^+HCl_2^-$. IR measurements show that this anion has the symmetrical linear structure Cl-H-Cl⁻. The anions HBr_2^- and HI_2^- have also been found in a few compounds. Mixed anions such as $HClBr^-$ and $HClI^-$ in the compounds $(n-C_4H_9)_4N^+HClBr^-$ and $(n-C_4H_9)_4N^+HClI^-$, are also known.

3.2.2. Acids, Bases, and Salts.

The characterization of substances as acids and bases in the higher hydrogen halides
follows the same scheme as in hydrogen fluoride. Since the acidity of the higher
hydrogen halides is also very strong (the acid strength too increases in the order HF
HCl < HBr < HI), the number of solvo acids is very small. Methyl- and trifluoro-
methylsulphonic acids and chlorosulphuric acid considerably increase the conductivity of liquid hydrogen chloride, but cannot be titrated conductimetrically as acid
Only HBr and HI appear to occur as weak solvo acids in liquid hydrogen chloride.
As in liquid hydrogen fluoride, ansolvo acids (halide ion acceptors) are more numer
Examples of these are the boron halides, just as boron trifluoride is an ansolvo acid
in liquid hydrogen fluoride:

$$BX_3 + 2\,HX \rightleftharpoons H_2X^+ + BX_4^-.$$

However, boron trichloride is only slightly ionized in liquid hydrogen chloride. The
phase diagram shows no compound formation. Tetrahalogenoborates, on the other
hand, are readily obtainable by reaction of boron trihalides with bases in the appro-
priate hydrogen halides. Difficulties can arise if the base was formed by protonation, since a solvent molecule may then be split off to form an adduct, as the case
e.g. with dimethyl sulphoxide as the base.

$$(CH_3)_2SOH^+HCl_2^- + BCl_3 \xrightarrow{\text{liq. HCl}} (CH_3)_2SOH^+BCl_4^- + HCl$$
$$\downarrow$$
$$(CH_3)_2SO \cdot BCl_3 + HCl.$$

In other cases, however, even bases of this type give salts, as is shown by the sharp
end-points obtained in conductimetric titrations.

$$2\,(CH_3)_4N^+HCl_2^- + B_2Cl_4 \xrightarrow{\text{liq. HCl}} [(CH_3)_4N^+]_2B_2Cl_6^{2-} + 2\,HCl.$$

For example, with boron trifluoride in liquid hydrogen chloride, salts of the acid
HBF_3Cl can be isolated. Boron trifluoride exhibits no acceptor properties in liquid
hydrogen bromide, while boron trichloride is solvolysed to boron tribromide, which
acts as an acid by functioning as a bromide acceptor in this solvent. Thus conductimetric titration of a solution of boron tribromide in liquid hydrogen bromide with
a sharp bend at a molar ratio of 1 : 1.
Of the halides of the group 4 elements, only tin tetrachloride has been found to
exhibit acidic properties in liquid hydrogen chloride, and tin tetrabromide in liquid
hydrogen bromide.

The halides of the group 5 elements behave very differently in the higher liquid hydrogen halides. Phosphorus trichloride and phosphorus trifluoride are neutral in liquid hydrogen chloride, whereas arsenic trifluoride is completely solvolysed. Phosphorus pentachloride and phosphorus pentabromide behave as bases, while phosphorus pentafluoride reacts with strong bases to form hexafluorophosphates:

$$2\,Me_4NCl + 3\,PF_5 \longrightarrow 2\,Me_4NPF_6 + PF_3Cl_2.$$

The following reaction occurs with PCl_5:

$$2\,PCl_5 + 3\,PF_5 \longrightarrow 2\,PCl_4{}^+PF_6{}^- + PF_3Cl_2.$$

Phosphorus pentafluoride is the strongest known acid in liquid hydrogen chloride. Arsenic pentafluoride is solvolysed in liquid hydrogen chloride with formation of tetrachloroarsonium hexafluoroarsenate:

$$2\,AsF_5 + 4\,HCl \longrightarrow AsCl_4{}^+AsF_6{}^- + 4\,HF.$$

The number of solvo and ansolvo bases in the higher hydrogen halides is again much greater than the number of acids. Solvo bases include salts of the hydrogen halides with large cations, whose lattice energies are sufficiently low to permit dissolution. Tetraalkylammonium halides are the strongest bases. All readily protonated systems, such as amines and compounds containing oxygen functions are comparatively rare. Solvo bases are also found among the halides of the group 5 elements; for example, phosphorus pentachloride reacts as a base:

$$PCl_5 + 2\,HCl \longrightarrow PCl_4{}^+ + HCl_2{}^-.$$

Salts of the cation $PCl_4{}^+$ can be prepared by neutralization with acids. Though phosphorus pentachloride exists as a salt $PCl_4{}^+PCl_6{}^-$ in the solid state, $PCl_6{}^-$ ions have not been detected in liquid hydrogen chloride. Another group of solvo bases consists of the carbon-halogen compounds whose carbon-halogen bonds are readily broken to form a carbonium ion. The ionization of triphenylmethyl chloride in liquid hydrogen chloride is easily recognizable from the deep colour of the solution. Protonations have been investigated with organic compounds of the elements of groups 4,5, and 6, various olefins, acetylenes, nitriles, azo compounds, and organic compounds containing oxygen. Triphenylamine and trimethylamine are strong bases in liquid hydrogen chloride. The nitrogen in hydrazobenzene is also protonated, but this is followed by rearrangement to form benzidine. Pyridine is an excellent base in all three solvents. Pyridinium tetrahalogenoborates are readily obtainable. Phosphine and triphenylphosphine are also protonated by liquid hydrogen chloride, bromide and iodide. Salts such as phosphonium tetrachloroborate and phosphonium tetrabro borate can be obtained from phosphine. Triphenylarsine is also a strong base in liqui hydrogen chloride; though its tetrachloroborate cannot be obtained because of addu formation:

$$(C_6H_5)_3As + 2 HCl \longrightarrow (C_6H_5)_3AsH^+ + HCl_2^-$$

$$(C_6H_5)_3AsH^+ + BCl_3 + HCl_2^- \longrightarrow (C_6H_5)_3AsH^+BCl_4^- + HCl$$
$$\downarrow$$
$$(C_6H_5)_3As \cdot BCl_3 + HCl.$$

Many ethers and thioethers give electrically conducting solutions in liquid hydrogen chloride. Alcohols are also protonated. Water and hydrogen sulphide, on the other hand, are insoluble in liquid hydrogen chloride; soluble basic compounds are obtained only on replacement of one or both hydrogen atoms by methyl or phenyl groups. Dimethyl sulphide is the strongest base of this group. However, tetrachloroborates cannot be isolated, since adducts are always formed with boron trichloride.

1,1-Diphenylethylene is protonated by liquid hydrogen chloride to form a carboniur ion:

$$(C_6H_5)_2C = CH_2 + 2 HCl \longrightarrow (C_6H_5)_2C^+ - CH_3 + HCl_2^-.$$

Similar reactions have been observed with styrene and α-methylstyrene. Carbonium ions are formed according to their stability, which increases in the order

$$(CH_3)_3C^+, (C_6H_5)(CH_3)HC^+, (C_6H_5)(CH_3)_2C^+ \ll (C_6H_5)_2CH_3C^+ < (C_6H_5)_3C^+$$

Doubly charged carbonium ions have also been found. The existence of the doubly charged ion $(C_6H_5)C^{2+} - CH_3$ can be detected by conductimetric titration of phenylacetylene with boron trichloride in liquid hydrogen chloride:

$$(C_6H_5)C \equiv CH + 4 HCl \longrightarrow (C_6H_5)C^{2+} - CH_3 + 2 HCl_2^-$$

$$(C_6H_5)C^{2+} - CH_3 + 2 HCl_2^- + 2 BCl_3 \longrightarrow (C_6H_5)C^{2+} - CH_3 + 2 BCl_4^- + \ ?$$

The oxygen in carbonyl, nitro, phosphoryl, sulphinyl, and sulphonyl groups also exhibit basic properties in liquid hydrogen chloride (this is the only solvent examine Apart from the aldehydes, most compounds of this type are strong bases, though reaction with boron trichloride nearly always yields adducts and not true salts. Ion pairs occur in solution in many cases, but on removal of the solvent they lose one molecule of hydrogen chloride to form adducts. Thus diphenylphosphoryl chloride reacts with boron trichloride in liquid hydrogen chloride as follows:

$$(C_6H_5)_2PClOH^+HCl_2^- + BCl_3 \longrightarrow (C_6H_5)_2PClOH^+BCl_4^- + HCl$$
$$\downarrow$$
$$(C_6H_5)_2PClO \cdot BCl_3 + HCl.$$

3.2.3. Solvolyses

The solvolysis of arsenic trifluoride and arsenic pentafluoride in liquid hydrogen chloride was mentioned in Section 3.2.2. This type of solvolysis, i.e. the replacement of the lighter halogen by the heavier, is particularly common in the higher hydrogen halides. The solvolysis of antimony trifluoride similarly yields antimony trichloride:

$$SbF_3 + 3\,HCl \longrightarrow SbCl_3 + 3\,HF.$$

Owing to the strength of the B-F bond, boron trifluoride is not solvolysed either in liquid hydrogen chloride or in liquid hydrogen bromide. However, boron trichloride gives boron tribromide in liquid hydrogen bromide. Both of these boron halides are converted into boron triiodide by liquid hydrogen iodide:

$$BCl_3 + 3\,HI \longrightarrow BI_3 + 3\,HCl.$$

Other solvolyses lead to the replacement of phenyl or hydroxyl groups halide ions. Thus triphenylmethanol is readily solvolysed by liquid hydrogen chloride:

$$(C_6H_5)_3COH + 3\,HCl \longrightarrow (C_6H_5)_3C^+HCl_2^- + H_3O^+Cl^-.$$

The driving force of this process is probably the insolubility of hydroxonium chloride and the high stability of the triphenylmethyl anion. Triphenyltin chloride is solvolysed with formation of benzene:

$$(C_6H_5)_3SnCl + HCl \longrightarrow (C_6H_5)_2SnCl_2 + C_6H_6.$$

3.2.4. Oxidation-Reduction Reactions

The higher hydrogen halides have also been used as solvents for redox reactions. Unlike in hydrogen fluoride, however, oxidations can proceed with liberation of the halogen. The strongest oxidizing agent in liquid hydrogen chloride that does not attack the solvent is chlorine. Iodine is not very soluble in liquid hydrogen chloride, and does not appear to act as an oxidizing agent in this solvent. The iodide ion is oxidized by chlorine to ICl_2^- and ICl_4^-; both stages can be recognized by conductimetric titration. The bromide ion is oxidized by chlorine to $BrCl_2^-$, and the iodide ion by bromine to IBr_2^-. Iodine monochloride reacts with the iodide ion in liquid hydrogen chloride to form iodine and ICl_2^-:

$$I^- + ICl + HCl \longrightarrow I_2\downarrow + HCl_2^-$$

$$ICl + HCl_2^- \longrightarrow HCl + ICl_2^-.$$

The second step in this process is not a redox reaction, but an acid-base reaction. Another example of the use of chlorine or bromine as an oxidizing agent is the reaction with phosphorus trichloride in liquid hydrogen chloride:

$$PCl_3 + Cl_2 + HCl \longrightarrow PCl_4^+ + HCl_2^-$$

$$PCl_3 + Br_2 + HCl \longrightarrow PCl_3Br^+ + HClBr^-.$$

On removal of the solvent, the ion PCl_3Br^+ decomposes, but the salt $PCl_3Br^+BCl_4^-$ can be isolated with boron trichloride.

3.3. Bibliography

H.H. Hyman u. J.J. Katz: Liquid Hydrogen Fluoride, in T.C. Waddington: Non Aqueous Solvent Systems. Academic Press, London, New York 1965, Chapter 2

L.F. Audrieth u. J. Kleinberg: Non-Aqueous Solvents. Wiley, New York 1953, Chapter 10

G. Jander: Die Chemie in wasserähnlichen Lösungsmitteln. Springer, Berlin, Göttingen Heidelberg 1949, chapter 2

J.H. Simons: Fluorocarbons and Their Production, in J.H. Simons: Fluorine Chemistry. Academic Press, New York 1950, Vol. 1, Chapter 12

A.F. Clifford, H.C. Beachell, W.M. Jack, A.G. Morris u. S. Kongpricha, J. inorg. nucl. Chem. *5*, 57, 71, 76 (1958); A.F. Clifford u. J. Sargent, J. Amer. chem. Soc. **79**, 4041 (1957); S. Kongpricha u. A.F. Clifford, J. inorg. nucl. Chem. **18**, 270 (1961)

F. Seel, Angew. Chem. **77**, 689 (1965)

M.E. Peach u. T.C. Waddington: The Higher Hydrogen Halides as Ionizing Solvents, in T.C. Waddington: Non-Aqueous Solvent Systems. Academic Press, London, New York 1965, Chapter 3

4. Sulphuric Acid and Fluorosulphuric Acid

4.1. Sulphuric Acid

Sulphuric acid, like liquid hydrogen fluoride, is a solvent of high acidity (H_O value = -12.08). This property together with its wide and convenient liquid range led to the early use of sulphuric acid as a solvent. Particular interest was shown in the behaviour of weak bases in sulphuric acid. Following the initial work by Hantzsch at the beginning of the present century, chemistry in this solvent was studied in particular by Hammett and Gillespie.

4.1.1. Physico-Chemical Properties of Sulphuric Acid

Table 13. Physico-chemical properties of sulphuric acid

Freezing point (°C)	10.371
Boiling pt. (°C)	290 - 317
Viscosity (c Poise)	24.54
Density (g/cm^3)	1.8269
Dielectric constant	100 (25 °C)
Specific conductivity (Ω^{-1}cm^{-1})	0.010439
Heat capacity (cal · deg.$^{-1}$ · g^{-1})	0.3373
Heat of fusion (cal · mol^{-1})	2560

As can be seen from the physico-chemical properties of sulphuric acid (Table 13), this solvent, like liquid hydrogen fluoride, is highly associated, i.e. strong hydrogen bonds exist between the molecules. The crystal has a sort of layer lattice, in which each sulphuric acid molecule is linked to four other molecules by hydrogen bonds. The structure of the liquid is probably similar to that of solid sulphuric acid. Its relatively high specific conductivity is due to a selfionization stronger than that of water, liquid ammonia, or liquid hydrogen fluoride:

$$2 \, H_2SO_4 \rightleftharpoons H_3SO_4^+ + HSO_4^-.$$

The ionic product K = $[H_3SO_4^+][HSO_4^-]$ is 2.7 x 10^{-4} at 25 °C, and is thus higher than that of any other self-ionizing liquid. However, an additional self-ionization beginning with the formation of water and sulphur trioxide also occurs:

$$H_2SO_4 \rightleftharpoons H_2O + SO_3$$

$$H_2O + H_2SO_4 \rightleftharpoons H_3O^+ + HSO_4^-$$

$$SO_3 + H_2SO_4 \rightleftharpoons H_2S_2O_7$$

$$H_2S_2O_7 + H_2SO_4 \rightleftharpoons H_3SO_4^+ + HS_2O_7^-.$$

The ionic product $K = [H_3O^+][HS_2O_7^-]$ is 5.1 x 10^{-5} at 25 °C and the dissociation constant of disulphuric acid

$$K = \frac{[H_3SO_4^+][HS_2O_7^-]}{[H_2S_2O_7]}$$

is 1.4 x 10^{-2} at 25 °C.

Metal cations (e.g. in solutions of metal hydrogen sulphates) dissolved in sulphuric acid have only a very low migration velocity (ion mobilities: $Na^+ \sim 3$, $Ba^{2+} \sim 2$, $H_3O^+ \sim 5$) because of the high viscosity of the solvent, and so contribute little to the transport of current. Most hydrogen sulphates have very similar specific conduc-

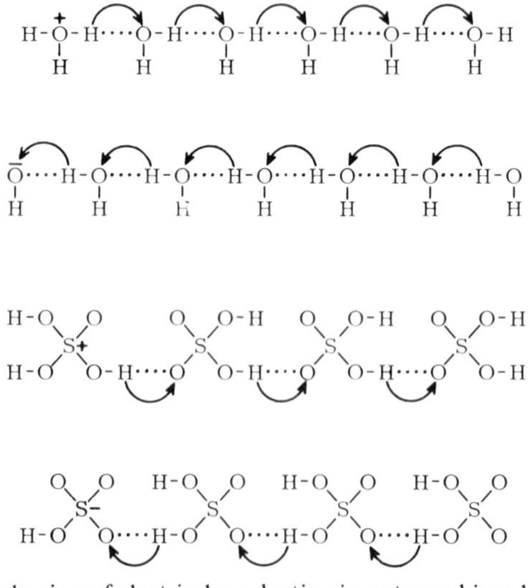

Mechanism of motion of the H_3O^+ ion (effective directic of motion: ⟶)

Mechanism of motion of the OH^- ion (effective direction of motion: ⟶)

Mechanism of motion of the $H_3SO_4^+$ ion (effective direct of motion: ⟶)

Mechanism of motion of the $H_2SO_4^-$ ion (effective directi of motion: ⟶)

Fig. 6. Mechanism of electrical conduction in water and in sulphuric acid (from H.H. Sisler: Chemistry in Non-Aqueous Solvents. Reinhold, New York 1961, p. 59).

tivities at low concentrations, indicating that the current is being transported mainly by the ion HSO_4^-. The migration velocities of the cation $H_3SO_4^+$ and of the anion HSO_4^- are in fact very high despite the high viscosity (ion mobilities $H_3SO_4^+$ 242, HSO_4^- 171). The transport numbers in a 0.56 M solution of $LiHSO_4$ are 0.013 for the cation and 0.987 for the anion, while those in a 0.17M solution of barium hydrogen sulphate are 0.009 and 0.991 respectively. These observations suggest that the solvent cations and anions of sulphuric acid do not migrate individually but by means of chains, similarly to the proton or the hydroxide ion in water (Fig. 6).

4.1.2. Solubility of Inorganic and Organic Substances in Sulphuric Acid; Solvates

Owing to its high dielectric constant, strong acidity, and strong hydrogen bonds, sulphuric acid is a good solvent for ionizing compounds, proton acceptors, and compounds that tend to form hydrogen bonds. The interpretation of the dissociation and protonation in terms of the solvent theory of acids and bases will be discussed in Section 4.1.3. For the present we shall confine ourselves to a brief statement of the facts.

Some qualitative solubility data for dissociating inorganic compounds (salts) are given in Table 14.

Table 14. Solubilities of some inorganic salts in sulphuric acid

Solubility	Solute
Abundantly soluble without noticeable solvolysis at room temperature	Li_2SO_4, K_2SO_4, Ag_2SO_4, $BaSO_4$, $NaNO_3$, KNO_3, $AgNO_3$, Na_2HAsO_4, $KSCN$, $K_4[Fe(CN)_6]$, $K_3[Fe(CN)_6]$, CaF_2, $Ca_3(PO_4)_2$
Slightly soluble	$CaCO_3$, Hg_2SO_4, $MgSO_4$, $ZnSO_4$, $FeSO_4$,
Insoluble	$AgCl$, $CuBr_2$, $CuSO_4$, $HgSO_4$, $PbSO_4$, $NiSO_4$, $AlCl_3$, $AlPO_4$, $Al_2(SO_4)_3$, $Fe_2(SO_4)_3$
Soluble with solvolysis and formation of soluble products	$NaCl$, $NaBr$, KCl, KBr, KI, $CaCl_2$
Soluble with solvolysis and formation of insoluble products	Na_2SiO_3, $Al(NO_3)_3$, $Fe(NO_3)_3$, $FeCl_3$, $Fe_4[Fe(CN)_6]_3$

Most salts are solvolysed on dissolution in sulphuric acid (cf. Section 4.1.4). Salts of metals whose sulphates are insoluble are either insoluble or are solvolysed to form the insoluble sulphates. The same is true of salts of acids that are insoluble in sulphuric acid in the free state. Conversely, salts of metals whose sulphates are readily soluble in sulphuric acid and salts whose acids are readily soluble in the free state are also readily soluble. Quantitative data are given in Table 15.

Table 15. Solubilities of some sulphates in sulphuric acid

Sulphate	Solubility at 25 °C (mol-%)	Sulphate	Solubility at 25 °C (mol-%)
Li_2SO_4	14.28	$ZnSO_4$	0.17
K_2SO_4	9.24	$FeSO_4$	0.17
$BaSO_4$	8.85	$Fe_2(SO_4)_3$	0.01
Na_2SO_4	5.28	$CuSO_4$	0.08
$CaSO_4$	5.16	Hg_2SO_4	0.78
Ag_2SO_4	9.11	$HgSO_4$	0.02

As in water, liquid ammonia, and liquid hydrogen fluoride, adducts are also formed in sulphuric acid. Most of those known are adducts of compounds of the solvent anion. The adducts $K_2SO_4 \cdot 3 H_2SO_4$, $K_2SO_4 \cdot H_2SO_4$, $(NH_4)_2 SO_4 \cdot 3 H_2SO_4$, $(NH_4)_2SO_4 \cdot H_2SO_4$, $Na_2SO_4 \cdot H_2SO_4$, and $Li_2SO_4 \cdot H_2SO_4$ are stable at melting point. Adducts that are unstable at the melting point include those of barium sulphate $(BaSO_4 \cdot 3 H_2SO_4)$, magnesium sulphate $(MgSO_4 \cdot 3 H_2SO_4)$, and silver sulphate $(Ag_2SO_4 \cdot 2 H_2SO_4$ and $Ag_2SO_4 \cdot H_2SO_4)$.

Many compounds are protonated and/or form hydrogen bonds on dissolving in anhydrous sulphuric acid. One of these compounds is water. The phase diagramm shows the existence of the following solvates having definite melting points: $4 H_2O \cdot H_2SO_4$, $2 H_2O \cdot H_2SO_4$ and $H_2O \cdot H_2SO_4$. Very dilute solutions of water in sulphuric acid, on the other hand, are completely protonated (as ion pairs $H_3O^+ HSO_4^-$), as is shown by conductivity measurements. IR and Raman spectra as well as cryoscopic measurements also show that the H_3O^+ ion is solvated, i.e. it is present as the $(H_3O.H_2SO_4)^+$ ion.

Organic compounds in particular are protonated on dissolution in sulphuric acid (cf. Section 4.3). These include amines, amides, nitriles, nitro compounds, esters, ketones, aldehydes, carboxylic acids and their anhydrides, sulphones, sulphoxides, and some aromatic hydrocarbons. Single protonation occurs in most cases, though polybasic compounds such as o-phenylenediamine and hexamethylenetetramine are protonated on all their basic centres as is shown by cryoscopic and conductivity measurements. Guanidine, which is a simple base in water, is also incompletely protonated in sulphuric acid, owing to considerable resonance stabilization of singly protonated species:

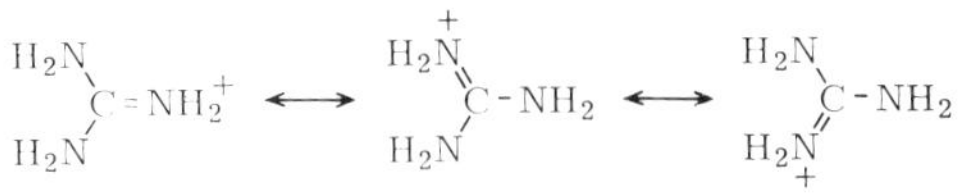

The protonation of amides and esters is followed by solvolysis. NMR measurements show that amides are protonated, not on the nitrogen, but on the oxygen. The solvolysis of amides in anhydrous sulphuric acid proceeds as follows:

$$RC(OH^+)NHR + 2 H_2SO_4 \longrightarrow RCO_2H_2^+ + RNH_3^+ + HS_2O_7^-$$

The rate-determining step is thought to be the formation of acyl ions.

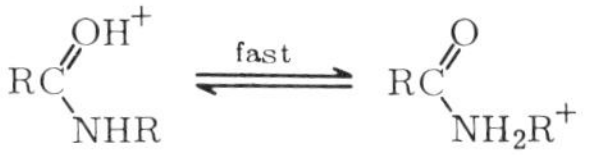

$$RCONH_2R^+ \xrightarrow{slow} RCO^+ + RNH_2$$

$$RNH_2 + H^+ \xrightarrow{fast} RNH_3^+$$

$$RCO^+ + H_2O \xrightarrow{fast} RCO_2H_2^+$$

Like the amides, most esters are solvolysed in sulphuric acid. However, the reaction rate is very low in many cases; thus methyl or ethyl benzoate can be recovered unchanged from cold sulphuric acid when the solution is poured onto ice. Two mechanisms are possible for the cleavage of esters, i.e. acyl oxygen and alkyl oxygen cleavage:

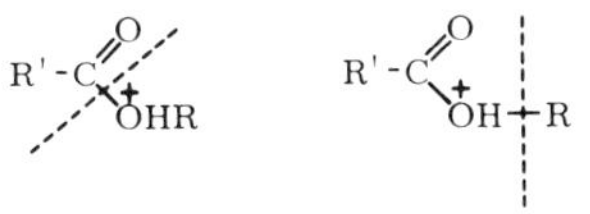

The mechanism followed in any particular case depends largely on the substituents R and R'. Electron-attracting substituents in R' hinder acyl oxygen cleavage, while substituents of this type in R facilitate it. Electron-attracting groups have the opposite effect on alkyl-oxygen cleavage. Thus methyl benzoate undergoes acyl oxygen cleavage while isopropyl benzoate exhibits alkyl oxygen cleavage. The overall equation for methyl benzoate is:

$$C_6H_5 - CO_2CH_3 + 2\,H_2SO_4 \longrightarrow C_6H_5 - CO_2H_2{}^+ + HSO_4^- + (CH_3)HSO_4.$$

The protonation of esters of mesitylic acid is followed by complete solvolysis to form stable acyl ions.

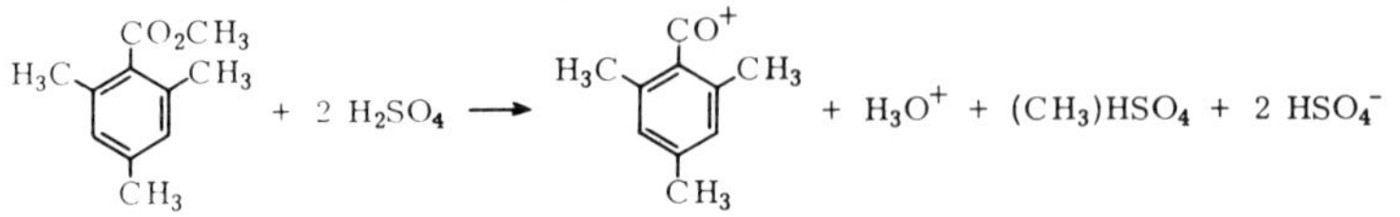

The esters of mesitylic acid can be hydrolysed in aqueous solution only with great difficulty. On the other hand, the free acid can be readily obtained by pouring solutions of its ester in sulphuric acid onto ice. The acid can in turn be readily esterified by pouring a solution in sulphuric acid into an alcohol.

Carboxylic acids are also normally protonated in sulphuric acid. Anhydrides may be formed in the case of dicarboxylic acids. Thus phthalic acid is singly protonated by sulphuric acid, whereas weak oleum gives the anhydride:

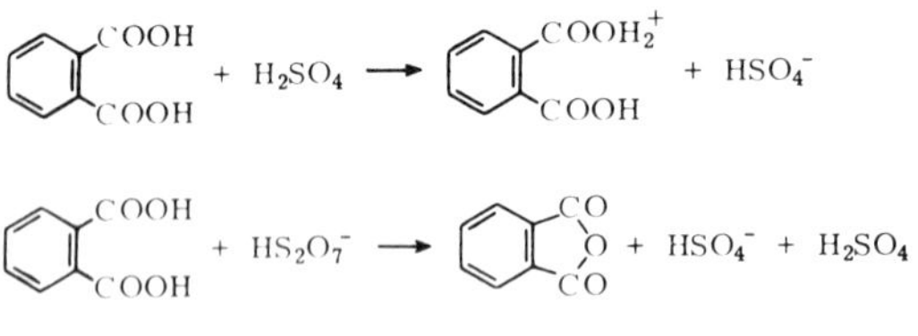

The equilibrium between the acid and its anhydride in anhydrous sulphuric acid is displaced toward the acid by water and toward the anhydride by sulphur trioxide. Many acids, including mesitylic acid, give stable acyl ions in sulphuric acid. Similar ion formation is observed with some aromatic keto-carboxylic acids, such as o-benzoyl-benzoic acid:

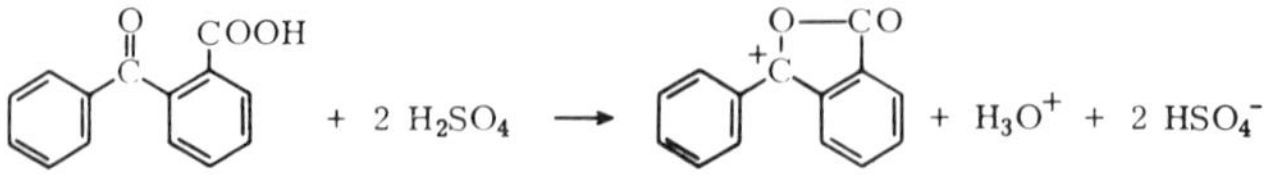

Some acids decompose in sulphuric acid to form carbon monoxide. These include formic acid, oxalic acid, and a-keto acids:

$$HCOOH + H_2SO_4 \longrightarrow CO + H_3O^+ + HSO_4^-$$

$$(COOH)_2 + H_2SO_4 \longrightarrow CO + CO_2 + H_3O^+ + HSO_4^-$$

$$C_6H_5-CO-COOH + H_2SO_4 \longrightarrow C_6H_5-COOH_2^+ + CO + HSO_4^-$$

Ethers are also protonated in sulphuric acid. However, the solutions are generally unstable, and alkyl hydrogen sulphates are formed by solvolysis.

$$R-O-R' + 3 H_2SO_4 \longrightarrow ROSO_3H + R'OSO_3H + H_3O^+ + HSO_4^- .$$

The rate-determining step is the cleavage of the conjugate acid with formation of the more stable of the possible carbonium ions. The protonated ether is assumed to react with sulphur trioxide (cf. Section 4.1.1.), the resulting oxonium compound then being split to form the alkyl ion:

$$R-\overset{+}{\underset{H}{O}}-R' + SO_3 \rightleftharpoons R-\overset{+}{\underset{SO_3H}{O}}-R' \longrightarrow R^+ + R'\,OSO_3H$$

The carbonium ion immediately reacts further with a hydrogen sulphate ion. Though the possibility of side reactions (e.g. sulphonations) make sulphuric acid less suitable as a reaction medium than liquid hydrogen fluoride, stable carbonium ions can be obtained in it, in some cases by dissociation and in other by protonation. Thus anthracene and 3,4-benzpyrene are protonated to give the following ions:

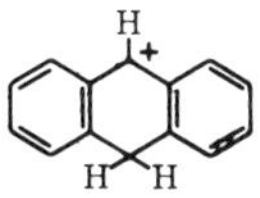

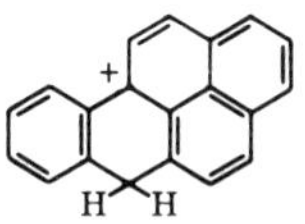

One of the best-known carbonium ions is the yellow triphenylmethyl ion, which is formed by dissociation when triphenylmethanol is dissolved in sulphuric acid:

$$(C_6H_5)_3COH + 2 H_2SO_4 \longrightarrow (C_6H_5)_3C^+ + H_3O^+ + HSO_4^- .$$

Para-substituted triphenylcarbinols ionize in a similar manner. The tris-p-dimethyl-amino derivative not only dissociates but is also protonated; however, it takes up

only two protons, so that the resulting carbonium ion is best formulated by a quinoi
structure:

The starting compounds are recovered unchanged on dilution of the sulphuric acid
with water. The dissolution of aliphatic polyenes or alcohols in sulphuric acid also
yields aliphatic carbonium ions, but these react further. Only tertiary alkyl carboniu
ions appear to have a certain stability.

In addition to the compounds that dissolve in sulphuric acid with dissociation or
protonation, i.e. that act as electrolytes, there are also a few substances that, though
sufficiently protophilic to form hydrogen bonds with the solvent, are not protonated
by it. Compounds of this type are non-electrolytes in sulphuric acid. They include
alkylsulphonyl fluorides and chlorides, sulphonyl chloride, picric acid, and some
aromatic polynitro compounds.

4.1.3. Acids, Bases, and Salts

According to the solvent theory of acids and bases, substances that increase the
concentration of $H_3SO_4^+$ ions in sulphuric acid are acids, while those that increase
the HSO_4^- ion concentration are bases. As expected from the strong acidity
of sulphuric acid, the number of true solvo acids, as in liquid hydrogen fluoride, is
very small. Most compounds that are solvo acids in the aqueous system behave as
ansolvo bases in sulphuric acid, since they are protonated and form the HSO_4^- ion,
which corresponds to the OH^- ion of aqueous chemistry, e.g.:

$$CH_3COOH + H_2SO_4 \rightleftharpoons CH_3COOH_2^+ + HSO_4^-$$

$$H_3PO_4 + H_2SO_4 \rightleftharpoons H_4PO_4^+ + HSO_4^-$$

$$HNO_3 + 2\,H_2SO_4 \rightleftharpoons NO_2^+ + H_3O^+ + 2\,HSO_4^- \text{ (Nitrating mixture)}$$

$$HF + 2\,H_2SO_4 \rightleftharpoons HSO_3F + H_3O^+ + HSO_4^- \cdot$$

Even perchloric acid, which is one of the strongest solvo acids in the aqueous system,
is only very slightly dissociated in sulphuric acid; the perchlorates are almost completely solvolysed in accordance with the equation

$$MClO_4 + H_2SO_4 \longrightarrow M^+ + HSO_4^- + HClO_4.$$

Fluorosulphuric acid is also only a very weak solvo acid in sulphuric acid. Disulphuric acid $H_2S_2O_7$ and the higher polysulphuric acids are weak acids in H_2SO_4. They are formed in accordance with equations given in Section 4.1.1. when the ansolvo acid sulphur trioxide is dissolved in sulphuric acid. The acid $H_2S_2O_7$ can be detected in various ways as well as the ions $HS_2O_7^-$ and $S_2O_7^{2-}$ in dilute oleum. The acid $H_2S_3O_{10}$ is also present in more concentrated oleum; $H_2S_4O_{13}$ and higher acids cannot be detected with certainty. The only fairly strong acid in the sulphuric acid system appears to be tetrakis (hydrogen sulphato) boric acid $HB(HSO_4)_4$. Boric acid and boron oxide are solvolysed on dissolving in sulphuric acid, and the anion of the complex sulphatoboric acid is formed.

$$H_3BO_3 + 6H_2SO_4 \longrightarrow B(HSO_4)_4^- + 3H_3O^+ + 2HSO_4^-$$

$$B_2O_3 + 9H_2SO_4 \longrightarrow 2B(HSO_4)_4^- + 3H_3O^+ + HSO_4^-.$$

The formation of the tetrakis (hydrogen sulphato) borate anion is thus analogous to the formation of the tetrafluoroborate or tetrahydroxyborate anions in liquid hydrogen fluoride or water. The acid is also formed on dissolution of boric acid or boron trioxide in oleum:

$$H_3BO_3 + 3H_2S_2O_7 \longrightarrow H_3SO_4^+ + B(HSO_4)_4^- + H_2SO_4$$

$$B_2O_3 + 3H_2S_2O_7 + 4H_2SO_4 \longrightarrow 2H_3SO_4^+ + 2B(HSO_4)_4^-.$$

However, concentrated solutions of this acid have a more complex composition; Raman spectroscopy shows the presence of the acids $H_2S_2O_7$ and $H_2S_3O_{10}$, which are probably formed by condensation reactions of the ion $B(HSO_4)_4^-$:

$$2\,B(HSO_4)_4^- \longrightarrow \left[(HSO_4)_2B{\overset{O}{\underset{O-S-O}{\diagup\diagdown}}}B(SO_4H)_2 \right]^{2-} + H_2S_2O_7 + H_2SO_4$$

In addition to these complex sulphatoboric acids, two complex acids of tin and lead are known. These are hexakis (hydrogen sulphato) stannic acid $H_2Sn(HSO_4)_6$ and hexakis (hydrogen sulphato) plumbic acid $H_2Pb(HSO_4)_6$, which can be obtained e.g. by solvolysis of the tetraacetates:

$$Sn(OAc)_4 + 10H_2SO_4 \longrightarrow H_2Sn(HSO_4)_6 + 4AcOH_2^+ + 4HSO_4^-$$

$$Pb(OAc)_4 + 10H_2SO_4 \longrightarrow H_2Pb(HSO_4)_6 + 4AcOH_2^+ + 4HSO_4^-.$$

The plumbic acid has the following dissociation constants:

$$H_2Pb(HSO_4)_6 + H_2SO_4 \rightleftharpoons H_3SO_4^+ + HPb(HSO_4)_6^-, K_1 = 1{,}2 \times 10^{-2}$$

$$HPb(HSO_4)_6^- + H_2SO_4 \rightleftharpoons H_3SO_4^+ \ Pb(HSO_4)_6^{2-}, K_2 = 1{,}8 \times 10^{-3}.$$

As expected, there are more bases (solvo and ansolvo bases) than acids in sulphuric acid. The simplest and at the same time the strongest bases are the solvo bases. These correspond to the hydroxides in water:

$$KHSO_4 \rightleftharpoons K^+ + HSO_4^-$$

$$KOH \rightleftharpoons K^+ + OH^-.$$

The secondary sulphates which corresponds to the metal oxides in aqueous chemistry are converted into hydrogen sulphates in sulphuric acid. This process is also similar to the behaviour of many oxides in water:

$$K_2SO_4 + H_2SO_4 \longrightarrow 2\,K^+ + 2\,HSO_4^-$$

$$K_2O + H_2O \longrightarrow 2\,K^+ + 2\,OH^-.$$

The number of ansolvo bases is very large, since there are many (particularly organic) compounds that can be protonated. In addition to acetic acid, these include all the other carboxylic acids; the relative base strengths of these compounds depend mainly on the nature of the substituents in the vicinity of the carboxyl group. Thus acetic acid is a strong base in sulphuric acid, while dichloroacetic acid is a weak base and tri-chloracetic acid is a nonelectrolyte. The organic ansolvo bases also include ketones esters, amines, amides, and phosphines:

$$R_2CO + H_2SO_4 \rightleftharpoons R_2COH^+ + HSO_4^-$$

$$RCO_2R' + H_2SO_4 \rightleftharpoons RCO_2R'H^+ + HSO_4^-$$

$$RNH_2 + H_2SO_4 \rightleftharpoons RNH_3^+ + HSO_4^-$$

$$RCONH_2 + H_2SO_4 \rightleftharpoons RC(OH)^+NH_2 + HSO_4^-$$

$$R_3P + H_2SO_4 \rightleftharpoons R_3PH^+ + HSO_4^-.$$

Triphenylamine and triphenylphosphine which are only very weak bases in water, are strongly basic in sulphuric acid. Some classes of substances, such as nitro compounds, sulphones, and nitriles are incompletely protonated. Molecules containing several functional groups can form cations carrying more than one charge. o-Phenyl-enediamine, for example is doubly protonated, amino acids take up protons on both the amino and the carboxyl groups, and hexamethylenetetramine takes up four protons. In many cases the substances can be recovered unchanged when the solution is poured onto ice. The dissolution of organic substances, which is considered here only in connection with their basic action, will be discussed in detail in Section 4.1.2. Water naturally also behaves as a strong base in sulphuric acid:

$$H_2O + H_2SO_4 \rightleftharpoons H_3O^+ + HSO_4^- .$$

Selenium dioxide is only a weak base

$$SeO_2 + H_2SO_4 \rightleftharpoons HSeO_2^+ + HSO_4^- .$$

Another interesting group of protonatable substances consists of a few metal carbonyls, in which the proton is bound to the central metal atom. An example of such a compound is triphenylphosphine-tetracarbonyliron;

$$Fe(CO)_4P(C_6H_5)_3 + H_2SO_4 \longrightarrow HFe^+(CO)_4P(C_6H_5)_3 + HSO_4^- .$$

$[P(C_6H_5)_3]_2Fe(CO)_3$ and the analogous arsenic compounds behave similarly. π-complexes such as $[\pi-C_5H_5Mo(CO)_3]_2$ and $[\pi-C_5H_5Fe(CO)_2]_2$ are also protonated on the metal by sulphuric acid. The formation of metal-hydrogen bonds can be readily seen from the nuclear magnetic resonance spectrum, which contains proton resonances at unusually high field strengths. Carbonyl-olefin-metal complexes do not form metal-hydrogen bonds in sulphuric acid, but give a sort of carbonium ion. For example, cycloheptatrienetricarbonyl iron gives the cation $C_7H_9Fe(CO)_3^+$:

Since acids and bases are known in sulphuric acid, salt formation by neutralization is also possible. An example of this type of reaction is the preparation of potassium tetrakis (hydrogen sulphato) borate:

$$KHSO_4 + HB(HSO_4)_4 \rightarrow KB(HSO_4)_4 + H_2SO_4$$

The end point of this reaction can be recognized conductimetrically. However, the salt does not precipitate out at the neutral point. To obtain crystalline products, it is necessary to use concentrated solutions of the complex boric acid. The potassium, sodium, or strontium salts isolated contain less sulphate than is to be expected from the equation. Salts with polymeric anions derived from the structure

(see above) are formed. Salts of disulphatoboric acid with the formulae $M^I B(SO_4)_2$ and $M^{II}[B(SO_4)_2]_2$ are obtained by reaction of the sulphate with boric acid sulphur trioxide in sulphuric acid.

$$(NH_4)_2SO_4 + 2\,B(OH)_3 + 6\,SO_3 \longrightarrow 2\,NH_4[B(SO_4)_2] + 3\,H_2SO_4.$$

When the barium salt having the corresponding composition is heated, sulphur trioxide is split off to form the salt $Ba[B_2O(SO_4)_3]$. Boron nitride reacts with sulphuric acid to form the ammonium salt:

$$BN + 2\,H_2SO_4 \longrightarrow NH_4[B(SO_4)_2].$$

All these compounds are very hygroscopic, hydrolyse readily and contain polymeric anions.

Salts of hexakis (hydrogen sulphato) stannic acid or their desolvated form, e.g.. $K_2Sn(SO_4)_3$ or $CaSn(SO_4)_3 \cdot 3\,H_2O$, are obtained on evaporation of solution of tin dioxide and the corresponding metal sulphate in sulphuric acid. The salts containing three molecules of water may also be regarded as compounds of tris (hydrogen sulphato) stannic acid, e.g. $Ag_2Sn(OH)_3(HSO_4)_3$. Salts of hexakis (hydrogen sulphato) plumbic acid or of trisulphatoplumbic acid e.g. $K_2Pb(SO_4)_3$ and $(NH_4)_2\,Pb(SO_4)_3$, are formed as precipitates on titration of the plumbic acid with the corresponding hydrogen sulphates in sulphuric acid.

4.1.4. Solvolyses

Owing to the strong acidity of sulphuric acid, nearly all salts in this solvent are salts of weak acids and strong bases, and are accordingly strongly solvolysed to form basic solutions:

$$NH_4ClO_4 + H_2SO_4 \longrightarrow NH_4^+ + HSO_4^- + HClO_4$$

$$Na_3PO_4 + 3\,H_2SO_4 \longrightarrow 3\,Na^+ + 3\,HSO_4^- + H_3PO_4.$$

Carbon compounds, i.e. organic compounds, frequently exhibit solvolysis following the protonation discussed in Sections 4.1.2. and 4.1.3. The reader is referred back to Section 4.1.2. We shall merely mention here the solvolysis of alcohols as a typical case:

$$C_2H_5OH + 2H_2SO_4 \longrightarrow C_2H_5OSO_3H + H_3O^+ + HSO_4^-.$$

Organosilicon, organotin, and organolead compounds exhibit particularly interesting solvolysis reactions in sulphuric acid. For example, hexamethyldisiloxane reacts in accordance with the equation:

$$[(CH_3)_3Si]_2O + 3H_2SO_4 \longrightarrow 2(CH_3)_3SiOSO_3H + H_3O^+ + HSO_4^-.$$

Trimethylethoxysilane also reacts with sulphuric acid to give trimethylsilicon hydrogen sulphate:

$$(CH_3)_3SiOC_2H_5 + 3H_2SO_4 \longrightarrow (CH_3)_3SiOSO_3H + C_2H_5SO_4H + H_3O^+ + HSO_4^-.$$

The solvolysis of pure silane has also been investigated. Thus tetraphenylsilane undergoes cleavage of all the Si–C bonds.

$$(C_6H_5)_4Si + 12H_2SO_4 \longrightarrow Si(HSO_4)_4 + 4C_6H_5SO_3H + 4H_3O^+ + 4HSO_4^-.$$

It must be concluded from conductimetric measurements, however, that silicon hydrogen sulphate polymerizes, e.g. to give a compound $Si_2O_3(HSO_4)_2$.

The solvolysis of similar tin compounds and sulphuric acids leads to the hexakis (hydrogen sulphato) stannic acid mentioned in Section 4.1.3. Tetramethyltin on the other hand, like the alkylsilicon compounds, is incompletely solvolysed.

$$(CH_3)_4Sn + H_2SO_4 \longrightarrow (CH_3)_3SnHSO_4 + CH_4.$$

The solvolysis of lead tetraacetate, which leads to hexakis (hydrogen sulphato) plumbic acid, has already been mentioned in Section 4.1.3.

The solvolysis products of nitrates, nitric acid, and nitrogen (V) oxide are interesting in that they are nitrating agents. Aromatic systems can be particularly readily nitrated in solutions of nitric acid or its salts in sulphuric acid. Raman spectroscopic studies show that the nitrating agent in this case is the nitronium ion NO_2^+, which is formed from potassium nitrate or nitric acid by solvolysis as follows:

$$KNO_3 + H_2SO_4 \longrightarrow HNO_3 + K^+ + HSO_4^-$$
$$HNO_3 + H_2SO_4 \longrightarrow H_2NO_3^+ + HSO_4^-$$
$$H_2NO_3^+ + H_2SO_4 \longrightarrow NO_2^+ + H_3O^+ + HSO_4^-.$$

Nitrogen pentoxide also gives nitronium ions in sulphuric acid:

$$N_2O_5 + 3\,H_2SO_4 \longrightarrow 2\,NO_2{}^+ + H_3O^+ + 3\,HSO_4{}^-.$$

Nitrites and dinitrogen trioxide are solvolysed in sulphuric acid to form the nitrosyl ion NO^+:

$$N_2O_3 + 3\,H_2SO_4 \longrightarrow 2\,NO^+ + H_3O^+ + 3\,HSO_4{}^-.$$

$$NaNO_2 + 3\,H_2SO_4 \longrightarrow NO^+ + H_3O^+ + Na^+ + 3\,HSO_4{}^-.$$

Dinitrogen tetroxide behaves as nitrosyl nitrate on solvolysis:

$$N_2O_4 + 3\,H_2SO_4 \longrightarrow NO_2{}^+ + NO^+ + H_3O^+ + 3\,HSO_4{}^-.$$

Hydrogen chloride and some other chlorides are solvolysed in sulphuric acid to form chlorosulphuric acid, which, like perchloric acid, is a very weak acid in sulphuric acid

$$HCl + 2\,H_2SO_4 \longrightarrow ClSO_3H + H_3O^+ + HSO_4{}^-$$

$$(CH_3)_3C_6H_2COCl + 3\,H_2SO_4 \longrightarrow (CH_3)_3C_6H_2CO^+ + ClSO_3H$$

$$+ H_3O^+ + 2\,HSO_4{}^-.$$

Iodic acid is solvolysed via a protonation stage to give iodyl hydrogen sulphate:

$$HIO_3 + 2\,H_2SO_4 \longrightarrow IO_2HSO_4 + H_3O^+ + HSO_4{}^-,$$

which probably exists in the solvated polymeric form containing I–O–I–O linkages. There is no evidence of the occurence of the cations $IO_2{}^+$ and $H_2IO_3{}^+$. In the presence of iodine, on the other hand, the solvolysis of iodic acid leads to iodosyl sulphate $(IO)_2SO_4$, which, like iodine (IV) oxide $I_2O_4 (= IO^+ \cdot IO_3{}^-)$, contains polymeric IO^+ ions. Solutions of iodosyl sulphate contain some free IO^+ or I^{3+} ions, as can be seen from the reactions with benzene derivatives. The I^{3+} cation is though to be formed in the following equilibrium:

$$IO^+ + 2\,H^+ \rightleftharpoons I^{3+} + H_2O.$$

In oleum, this equilibrium is displaced in favour of the I^{3+} cation, and the yellow iodosyl sulphate is converted into the white iodine sulphate having the composition $I_2(SO_4)_3 \cdot H_2SO_4$. This compound may be regarded as a partially desolvated iodine trihydrogen sulphate $I(SO_4)(HSO_4)$. It probably has the following polymeric structure:

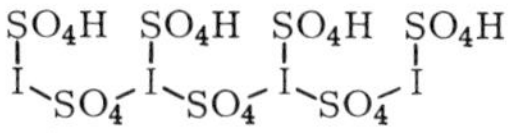

$$\begin{array}{cccc} SO_4H & SO_4H & SO_4H & SO_4H \\ | & | & | & | \\ I \diagdown {}_{SO_4} \diagdown {}^{I} \diagdown {}_{SO_4} \diagdown {}^{I} \diagdown {}_{SO_4} \diagdown {}^{I} \end{array}$$

Established views on the mechanism of solvolysis of iodic acid are based on cryoscopic and conductivity measurements. I^+ cations should be expected to occur at a molar ratio iodine: iodic acid = 2:

$$2\,I_2 + HJO_3 + 8\,H_2SO_4 \rightleftharpoons 5\,I^+ + 3\,H_3O^+ + 8\,HSO_4^-$$

$$2\,I_2 + HJO_3 + 8\,H_2SO_4 \rightleftharpoons 5\,IHSO_4 + 3\,H_3O^+ + 3\,HSO_4^-.$$

This reaction probably occurs in dilute solutions. In more concentrated solutions, on the other hand, the I^+ cation is disproportionated as follows:

$$4\,I^+ + H_2O + 2\,HSO_4^- \rightleftharpoons I_3^+ + IOHSO_4 + HSO_4^-$$

to form I_3^+ and iodosyl cations; the overall equation in this case is therefore:

$$8\,I_2 + 4\,HJO_3 + 17\,H_2SO_4 \rightleftharpoons 5\,I_3^+ + 7\,H_3O^+ + 5\,IOHSO_4 + 12\,HSO_4^-$$

This result is confirmed by optical studies, which show that the blue I^+ and the brown I_3^+ ions are present together at a molar ratio iodine : iodic acid = 2 in dilute solution. The spectrum of the I^+ ion is known, since this ion occurs alone on dissolution of iodine in sulphuric acid.

The following reaction takes place at a molar ratio iodine: iodic acid = 7

$$7\,I_2 + HJO_3 + 8\,H_2SO_4 \rightleftharpoons 5\,I_3^+ + 3\,H_3O^+ + 8\,HSO_4^-.$$

At molar ratios greater than 7, I_5^+ ions are formed to an increasing extent:

$$I_3^+ + I_2 \rightleftharpoons I_5^+.$$

4.2. Fluorosulphuric Acid

Fluorosulphuric acid exhibits the strongest acidity of all solvents. The anhydrous solvent is obtained from sulphur trioxide and hydrogen fluoride. Physico-chemical comparison with sulphuric acid shows difference above all in the melting point $(HSO_3F\!:\!89.\,0^\circ\,C;\,H_2SO_4 + 10.37\,^\circ C)$ and in the viscosity (1.56 as compared with 24.54 centipoise at 25 $^\circ$C). Fluorosulphuric acid evidently contains fewer hydrogen bonds than sulphuric acid. The specific conductivity of fluorosulphuric acid is lower by a factor of 100 than that of sulphuric acid. The mobility of the SO_3F^- and $H_2SO_3F^+$

ions is abnormally high in comparison with the mobilities of other ions, this can be interpreted by a proton transfer mechanism similar to that in sulphuric acid. Corresponding to the self-ionization of fluorosulphuric acid:

$$2\ HSO_3F \rightleftharpoons H_2SO_3F^+ + SO_3F^-$$

acids increase the concentration of the fluorosulphuric acid acidium ion, while bases increase the concentration of the fluorosulphonate ion. Since perchloric acid and sulphuric acid are proton acceptors in fluorosulphuric acid, it must be concluded that true solvo acids are non-existent in anhydrous fluorosulphuric acid. Thus any acids in this solvent must be ansolvo acids, i.e. acids that increase the $H_2SO_3F^+$ ion concentration by binding SO_3F^- ions. Examples of such substances are SbF_5, AuF_3, BF_3, TaF_5, PtF_4, and SO_3.

The number of solvo bases on the other hand is considerable, since many ionizing alkali and alkaline earth metal fluorosulphates are known. Ansolvo bases, as in sulphuric acid, are all protonatable substances. Even compounds such as AsF_3 and SbF_3 are ansolvo bases; they react with fluorosulphuric acid in accordance with the following scheme:

$$AsF_3 + HSO_3F \rightleftharpoons HAsF_3{}^+ + SO_3F^-$$

$$AsF_3 + HSO_3F \rightleftharpoons AsF_2(SO_3F) + HF$$

$$AsF_2(SO_3F) \rightleftharpoons AsF_2{}^+ + SO_3F^-.$$

Owing to the existence of acids and bases in fluorosulphuric acid, neutralization reactions are also possible. The acidic or basic character of a substance is shown directly by conductimetric titration with potassium fluorosulphate (base) or antimony pentafluoride (acid). For example, when potassium fluorosulphate is added to a base, the conductivity increases linearly, whereas on addition of the salt to an acid, the conductivity passes through a minimum owing to the neutralization of the fluorosulphate.

Because of the strong acidity of fluorosulphuric acid, the dissociation of compounds in this solvent is frequently associated with solvolysis. For example, potassium sulphate is completely solvolysed:

$$K_2SO_4 + 2\ HSO_3F \longrightarrow 2\ K^+ + 2\ SO_3F^- + H_2SO_4.$$

Anhydrous fluorosulphuric acid can thus be compared to some degrees with anhydrous sulphuric acid, though chemistry in fluorosulphuric acid exhibits less variety.

4.3. Bibliography

R.J. Gillespie u. E.A. Robinson: Sulfuric Acid, in T.C. Waddington: Non-Aqueous
    Solvent Systems. Academic Press, London, New York 1965, Chapter 4
L.F. Audrieth u. J. Kleinberg: Non-Aqueous Solvents. Wiley, New York 1953,
    Chapter 9
H.H. Sisler: Chemistry in Non-Aqueous Solvents. Reinhold, New York 1961
A. Engelbrecht, Angew. Chem. **77**, 695 (1965)

5. Acetic Acid (Glacial Acetic Acid)

5.1. Physico-Chemical Properties of Acetic Acid

Having discussed the solvents of strong acidity, we shall now consider one of less acidity, which has played an important part as a solvent, particularly in organic chemistry, for some time. Anhydrous acetic acid, which is easy to prepare and presents no difficulties in use, has a number of interesting physico-chemical properties, as can be seen from Table 16.

Table 16. Physico-chemical properties of anhydrous acetic acid

Melting point ($^{\circ}$C)	16.6
Boiling pt. ($^{\circ}$C)	118.5
Dielectric constant	6.29 (at 19 $^{\circ}$C)
Dipole moment	0
Electrical conductivity ($\Omega^{-1}cm^{-1}$)	0.4×10^{-8} (at 25 $^{\circ}$C)
Ionic product	3.5×10^{-15}(at 25 $^{\circ}$C)
Viscosity (c Poise)	1.13 (at 25 $^{\circ}$C)

The reason why the dipole moment is zero is that acetic acid is not monomeric, but dimeric:

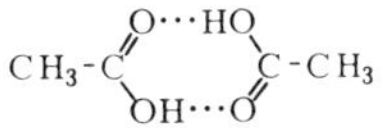

5.2. Solubilities in Acetic Acid, Solvates

Owing to its low dielectric constant, acetic acid dissolves covalent compounds better than ionic. Ionic compounds in solution should not have high degrees of dissociation, and should tend to form ion pairs. Owing to the hydrogen bonds between the solvent molecules, substances that can form hydrogen bonds and that can break down the solvent dimers to form solvates containing hydrogen bonds (organic substances and many acetates) should be particularly soluble. These general solution characteristics deduced from the physico-chemical properties of anhydrous acetic acid are found to be correct on the whole. For example, $HClO_4$ which is a strong acid in water, is only weakly acidic in acetic acid (dissociation constant 10^{-6}) because of the acidity of this solvent. It is often difficult, however, to predict solubilities and degrees of dissociation in acetic acid in individual cases. Some qualitative indications on the solubilities of inorganic compounds are given in Table 17.

Table 17. Solubilities of some inorganic compounds in anhydrous acetic acid

Solubility	Solute
Abundantly soluble	Liac, LiCl, LiBr, LiI, $LiNO_3$,
	Naac, Kac, KCN, NH_4ac, NH_4SCN,
	NH_4NO_3, $Cu(NO_3)_2$, $AgClO_4$,
	$Ca(NO_3)_2$, $Ba(ac)_2$, BaI_2, $ZnCl_2$,
	ZnI_2; $Cd(ac)_2$, $Pb(ac)_2$, $AsBr_3$
	$SbCl_3$, $SbBr_3$, $FeCl_3$
Fairly abundantly soluble	NaBr, $NaNO_3$, KCl, KBr, KI, KNO_3,
	$KClO_3$, NH_4Cl, NH_4Br, NH_4I, $Cu(ac)_2$,
	Agac, $AgNO_3$, $MgCl_2$, $Ca(ac)_2$, $CaCl_2$,
	$Zn(ac)_2$, $HgCl_2$, $HgBr_2$, HgI_2, $AlCl_3$, $CoCl_2$
Insoluble	NaCl, silver halides, AgCN, AgSCN,
	$CaCO_3$, $BaCl_2$, $Ba(NO_3)_2$, Hg_2Cl_2, CdI_2,
	$PbCl_2$, PbI_2, heavy metal sulphides,
	all sulphates and phosphates.

These solubilities in glacial acetic acid, some of which differ considerably from those in water, offer new possibilities for the separation of inorganic ions by precipitation reactions. Table 18 shows the course of a separation of cations (not yet complete) in glacial acetic acid, which illustrates the use of ionizing solvents in analytical chemistry.

Table 18. Scheme of cation separation in glacial acetic acid

Sample contains the anhydrous acetates of	Ag, Tl, Hg, Bi, Pb, Cu, Cd, Cr, Al, Ni, Zn, Ba, Sr, Mg, NH_4
Residue on dissolution in glacial acetic acid	Cr, Al
Addition of KSCN in glacial acetic acid to precipitate the thio-cyanates of	Ag, Tl, (Hg^*),Pb, Cu, Cd, Ni
Addition of H_2SO_4 in glacial acetic acid to precipitate the sulphates of	Bi, Zn, Ba, Sr, Mg, NH_4

The thiocyanate precipitate is soluble in water, and the sulphate precipitate in aqueous nitric acid. Heavy metal ions can be precipitated as sulphides by passage of hydrogen sulphide into solutions of their salts in glacial acetic acid, and barium sulphate can be obtained as a flocculent precipitate by addition of sulphuric acid to a solution of barium iodide in acetic acid.

The disadvantage of precipitation reactions in glacial acetic acid is that in many cases the precipitation is not quantitative because of the low degree of dissociation of the salts. However, quantitative determinations are possible in this solvent. For example, nitrate can be determined by conductimetric precipitation titration with barium acetate, and sodium by precipitation titration with sulphuric acid or oxalic acid dihydrate. Nickel can be separated from cobalt by precipitation of nickel thiocyanate, while cobalt gives the soluble, deep blue $[Co(SCN)_4]^{2-}$.

5.3. Acids, Bases, and Salts

The acid-base relationship in glacial acetic acid fit the solvent theory of acids and bases if considered in the light of the self-ionization of acetic acid:

$$2\,CH_3COOH \rightleftharpoons CH_3COOH_2^+ + CH_3COO^-.$$

---

*) Precipitates out as $Hg(SCN)_2$ and dissolves with excess reagent as $[Hg(SCN)_4]^{2-}$

Acetates are thus solvo bases, and proton donors stronger than acetic acid are solvo acids.

The solvo acids in glacial acetic acid include the strongest acids of the aqueous system. From the differentiating action of the solvent follows that acid strength decreases in the order:

$$HClO_4 > HBr > H_2SO_4 > HCl > HNO_3.$$

Trichloroacetic acid is also a solvo acid in this solvent. Even perchloric acid is much weaker in acetic acid than in water; on the other hand, its acidity is stronger because of the stronger protonactivity of the acetonium-ion $CH_3C(OH)_2^+$ compared to that of the hydronium-ion H_3O^+.

By definition, all soluble acetates are solvo bases. Potassium and ammonium acetates are among the strongest of these, though they are only slightly dissociated.

Ansolvo bases, which increase the acetate ion concentration by proton capture, are also known. These are all relatively strong bases, owing to the levelling action of the solvent on base strengths. One such base is guanidine. Potentiometric titration of guanidinium acetate with perchloric acid gives a steep neutralization curve similar to that obtained on titration of a strong acid with a strong base in water. Glacial acetic acid is therefore suitable for use as a solvent for the quantitative determination of various substances that are very weak bases in water. For example, amino acids, polypeptides, primary, secondary, and tertiary amines, and epoxides can be titrated with perchloric acid in acetic acid. The titration can be followed potentiometrically with a glass measuring electrode and a calomel reference electrode or with indicators. Another typical neutralization is the reaction of sodium acetate with perchloric acid:

$$CH_3COOH_2^+ + ClO_4^- + Na^+ + CH_3COO^- \longrightarrow Na^+ + ClO_4^- + 2\,CH_3COOH.$$

5.4. Solvolysis, Amphoterism

Examples of solvolysis and amphoterism are also known. Perchlorates, halides, and nitrates are solvolysed in glacial acetic acid with formation of the corresponding acids. Solutions of silver perchlorate and perchlorates of divalent cations give the strongest acidic reactions. An example of amphoteric behaviour is the solubility of zinc acetate in HCl/glacial acetic acid:

$$(CH_3COO)_2Zn + 2\,HCl \longrightarrow ZnCl_2 + 2\,CH_3COOH,$$

as well as in sodium acetate solutions:

$$(CH_3COO)_2Zn + 2CH_3COONa \longrightarrow Na_2[Zn(CH_3COO)_4].$$

This reaction is comparable to the formation of zincate in sodium hydroxide solution. Similarly, copper (II) acetate is soluble both in acid solutions and in potassium or ammonium acetate solutions.

5.5. Bibliography

L.F. Audrieth u. J. Kleinberg: Non-Aqueous Solvents. Eiley, New York 1953, Chapter8

H.H. Sisler: Chemistry in Non-Aqueous Solvents. Reinhold, New York 1961

K. Heymann u. H. Klaus: Chemie in wasserfreier Essigsäure, in G. Jander, H. Spandau u. C.C. Addison:Chemie in nichtwässrigen ionisierenden Lösungsmitteln. Vieweg, Braunschweig 1963,Vol. IV, Chapter 1

B. Sansoni u. R. Stolz, Angew. Chem. **75**, 418 (1963)

W. Huber: Titrationen in nichtwässerigen Lösungsmitteln, in F. Hecht, R. Kaiser u. H. Kriegsmann: Methoden der Analyse in der Chemie. Akademische Verlagsgesellschaft, Frankfurt 1964, Vol. 1

6. Liquid Hydrogen Sulphide and Liquid Hydrogen Cyanide

6.1. Liquid Hydrogen Sulphide

6.1.1. Physico-Chemical Properties of Liquid Hydrogen Sulphide

Since we have already discussed two non-aqueous ionizing solvents, i.e. liquid ammonia and liquid hydrogen fluoride, that are particularly close to water from the point of view of the periodic system, it is of interest now to consider hydrogen sulphide, which is also very close to water. We saw that ammonia exhibits basizity in comparison with water, while hydrogen fluoride shows acidity. Both have relatively high dielectric constants. Let us now consider the properties of hydrogen sulphide. This compound is certainly similar to water in many respects. It shows also (like liquid hydrogen cyanide) a very weak acidity, which resembles the organic solvents in several respects. The properties of hydrogen sulphide and water are compared in Table 19.

Table 19. Physico-chemical properties of hydrogen sulphide and water

	H_2S	H_2O
Molecular weight	34.09	18.016
Melting point (°C)	−85.5	0
Boiling point (°C)	−60.4	100
Density at boiling pt. (g/cm³)	0.950 (at −61 °C)	0.958 (at 100 °C)
Molar volume at boiling pt. (cm³)	35.9	18.8
Heat of formation of the gaseous compound (kcal/mol)	+5.3	+57.8
Dielectric constant	10.2 (at −60 °C)	81 (at 18 °C)
Electrical conductance (Ω^{-1})	3.7×10^{-11} (at −78 °C)	

Liquid hydrogen sulphide is a colourless, highly mobile liquid, which is practically unassociated, as can be seen from the melting and boiling points.

6.1.2. Solubilities in Liquid Hydrogen Sulphide

The relatively low dielectric constants leads to a relatively poor solvent power for inorganic compounds. Thus simple salts having a typical ionic lattice such as sodium fluoride or potassium chloride are insoluble in H_2S. Ionic compounds having large cations and anions are more soluble, owing to their smaller lattice forces. These include in particular alkyl- or aryl-substituted ammonium salts, e.g. $[C_2H_5)_4N]Cl$ or $[(CH_3)_3NH]$ (SH). The solubility of many salts is also improved when the size of the cation is increased by complex formation. For example, hexakis (diethylamine) nicke acetate $\{Ni[NH(C_2H_5)_2]_6\}$ $(CH_3COO)_2$ is more soluble than simple nickel acetate. The complex formation can be carried out in liquid hydrogen sulphide itself, and can be followed conductimetrically:

$$6\,[(C_2H_5)_2NH_2]SH + Cd(CH_3COO)_2 \longrightarrow \{Cd[NH(C_2H_5)_2]_6\} \ (CH_3COO)_2 + 6\,H_2$$

Like most simple salts, the hydrogen sulphides and sulphides of metals, including those of the alkaline metals, are insoluble. On the other hand, compounds having covalent bonds such as hydrogen chloride, hydrogen bromide, or acetic, trichloro-acetic, or sulphuric acid, which do not react in the anhydrous state with the anhy-drous solvent at low temperatures, are soluble. Substances that crystallize in molecula lattices, e.g. aluminium chloride, aluminium bromide, aluminium iodide, phosphorus (III) bromide, phosphorus (V) chloride, arsenic (III) chloride, antimony (III) chloride, antimony (V) chloride, bismuth chloride, iodine (III) chloride and iron (III) chlo-ride, are also soluble. However, some of these react with the solvent.

The behaviour of pure metals towards hydrogen sulphide is also of interest, since it leads to another electromotive series of the metallic elements that is different from that in water. Thus dissolution with liberation of hydrogen is observed not only with the alkali metals:

$$2\,Na + 2\,H_2S \longrightarrow 2\,NaHS + H_2\uparrow,$$

but also with copper, silver, and mercury which are noble metals in water:

$$2\,Ag + H_2S \longrightarrow Ag_2S + H_2\uparrow$$

$$2\,Cu + H_2S \longrightarrow Cu_2S + H_2\uparrow.$$

The alkali metals also exhibit polysulphide formation, corresponding to the forma-tion of peroxides:

$$2\,KHS \longrightarrow K_2S_2 + H_2\uparrow.$$

The reason for the difference in the electromotive series in liquid hydrogen sulphide and in water is the high affinity of hydrogen for sulphur. The heats of formation of the sulphides of silver, mercury, and copper are greater than that of hydrogen sulphide These metals consequently cannot be noble metals in liquid hydrogen sulphide.

Liquid hydrogen sulphide is an excellent solvent for organic substances. It dissolves e.g. hydrocarbons, halides, many acids, esters, acid chlorides, nitro and amino compounds, nitriles, alcohols, aldehydes, and ketones.

6.1.3. Acids, Bases, and Salts

According to the weak self-ionization of hydrogen sulphide

$$2\,H_2S \rightleftharpoons H_3S^+ + HS^-,$$

all compounds that increase the concentration of H_3S^+ ions are acids, and compounds that increase the concentration of hydrogen sulphide ions are bases. Solvo acids in the aqueous system are also solvo acids in hydrogen sulphide. These include hydrogen chloride and hydrogen bromide, as well as sulphuric, acetic, and trichloroacetic acids. Owing to the acidity (though only slight) of liquid hydrogen sulphide, all these compounds dissociate only to a very small extent. In the case of acetic acid, substitution to form trichloroacetic acid results in only a very small increase in the electrical conductivity of the solution.

A few halides of the elements of group 5 must be regarded as ansolvo acids because of the relatively high conductivity of their solutions. These include antimony trichloride, which forms a complex (hydrogen sulphide) chloroantimonic acid in liquid hydrogen sulphide:

$$SbCl_3 + 3\,H_2S \rightleftharpoons H_3[SbCl_3(SH)_3] \rightleftharpoons 3\,H^+ + [SbCl_3(SH)_3]^{3-}.$$

The high conductivity of the solutions is explained by the considerable dissociation of this acid; and could not be accounted for by hydrogen chloride formed by solvolysis. Solutions of arsenic (III) chloride and iron (III) chloride also have high conductivities.

All hydrogen sulphides should theoretically be solvo bases. Since most metal hydrogen sulphides and metal sulphides are insoluble in liquid hydrogen sulphide, however, practically the only compounds than can be used as bases are alkylamines and arylamines (ansolvo bases)

$$(C_2H_5)_3N + H_2S \rightleftharpoons [(C_2H_5)_3NH](SH) \rightleftharpoons [(C_2H_5)_3NH]^+ + (SH)^-.$$

Diisobutylamine, pyridine, piperidine, quinoline, and nicotine also form solutions with relatively high conductivities in liquid hydrogen sulphide. The dissociation of the hydrogen sulphides appears to depend on the size of the cations; thus the equivalent conductivity of triisobutylamine is greater than that of tripropylamine. In both cases, however. the equivalent conductivity decreases instead of increasing with increasing dilution. Attempts have been made to explain this remarkable behaviour by the assumption that not only does the addition product formed from the amine and the hydrogen sulphide exist in equilibrium with its dissociation products, but it is also associated with further amine, which can again ionize. The dissolution of amine in liquid hydrogen sulphide is frequently accompanied by considerable evolution of heat.

Neutralizations can also be carried out in liquid hydrogen sulphide by the action of the above acids and bases with one another, and can be detected sometimes prepara- tively and sometimes conductimetrically or potentiometrically. For example, sodium hydrogen sulphide, which is insoluble in liquid hydrogen sulphide, reacts with hy- drogen chloride to form the insoluble sodium chloride:

$$NaSH + HCl \longrightarrow H_2S + NaCl.$$

Neutralization reactions with the suspended alkali metal hydrogen sulphides are naturally difficult to follow by conductivity measurements. On the other hand, ti- tration of the soluble triethylammonium hydrogen sulphide with hydrogen chloride gives a conductivity curve a distinct salient point at a molar ratio of 1 : 1, which in dicates the formation of triethylammonium chloride:

$$[(C_2H_5)_3NH](SH) + HCl \longrightarrow H_2S + [(C_2H_5)_3NH]Cl.$$

Titration of triethylammonium hydrogen sulphide with sulphuric acid gives a different type of conductivity curve. The conductivity passes through a maximum at a molar ratio sulphuric acid: hydrogen sulphide = 0.25:1, and at a molar ratio of 0.5:1 it reaches a lower value, which remains practically constant on further addition of sulphuric acid. The interpretation of this curve is that before the triethylammonium sulphate (sulphuric acid: hydrogen sulphide = 0.5:1) stage is reached, a still more strongly dissociated basic sulphate (molar ratio 0.25:1) is formed.

$$4\,[(C_2H_5)_3NH]\,(SH) + H_2SO_4 \longrightarrow 4\,H_2S + [(C_2H_5)_3N \cdot (C_2H_5)_3NH]_2(SO_4$$

$$[(C_2H_5)_3N \cdot (C_2H_5)_3NH](SO_4) + H_2SO_4 \longrightarrow 2\,[(C_2H_5)_3NH]_2SO_4.$$

Yet another shape of curve is obtained on neutralization of triethylammonium hydrogen sulphide with trichloroacetic acid. In this case the expected salt is formed first:

$$[(C_2H_5)_3NH](SH) + CCl_3COOH \longrightarrow H_2S + [(C_2H_5)_3NH](OOCCCl_3).$$

The conductivity passes through a minimum at the equivalence point, but increases again on further addition of trichloroacetic acid, though the acid alone exhibits practically no conductivity in liquid hydrogen sulphide. This increase in conductivity is thought to be connected with the addition of trichloroacetic acid to the normal salt to form the adduct

$$[(C_2H_5)_3NH](OOCCCl_3) \cdot CCl_3COOH.$$

The result is a further salient point in the conductivity curve at a molar ratio trichloroacetic acid: triethylammonium hydrogen sulphide = 2:1.
The change in the H_3S^+ and SH^- ion concentrations can also be detected by means of colour indicators. For example, methyl red is red in the acid range and orange-yellow in the basic range in liquid hydrogen sulphide. Tropaeolin 00 is violet in acid solution and yellow in basic solution (red and yellow respectively in water).

6.1.4. Solvation, Solvolysis, and Amphoterism

While the description "water-like solvent" is justified for liquid hydrogen sulphide by the data presented already, the similarity to water becomes even clearer on consideration of solvolysis and amphoterism. Though liquid sulphide, owing to its low dielectric constant, is very much closer to the organic solvents than to water, solvolysis (thiohydrolyses) are known. Many compounds of transition metals (silver, cadmium, mercury) are solvolysed to form the sulphides; this is probably mainly due to their slight solubility. Other transition metal compounds such as copper, cadmium, cobalt, and nickel chlorides, on the other hand, are so insoluble in liquid hydrogen sulphide that they remain unchanged. Aluminium chloride fails to exhibit thiohydrolysis. Of the chlorides of the group 4 elements, only tin tetrachloride exhibits slight solvolysis (formation of tin disulphide), whereas carbon tetrachloride and silicon tetrachloride dissolve without appreciable reaction. Lead tetrachloride is reduced to lead (II) chloride. Parallels with the aqueous system are found in particular with the chlorides of the group 5 elements. Thus phosphorus pentachloride is immediately thiohydrolysed to form phosphorus sulphide chloride $PSCl_3$, Phosphorus trichloride does not react with hydrogen sulphide at low temperatures, but immediately gives phosphorus (III) sulphide at room temperature. Arsenic trichloride is converted into arsenic trisulphide at all temperatures, whereas antimony trichloride reacts only with gaseous hydrogen sulphide to give a lemon-yellow compound having the composition $SbSCl \cdot 7 H_2S$. Antimony pentachloride reacts in the same way as phosphorus pentachloride, with formation of antimony sulphide chloride $SbSCl_3$.

It is found in the case of arsenic trichloride that, as in the aqueous system, solvolysis is suppressed by an increase in the hydrogen ion concentration. No trisulphide is formed from arsenic trichloride in a solution of hydrogen chloride and liquid hydrogen sulphide. Conversely, the solvolysis of antimony trichloride can be forced by addition of triethylamine to increase the hydrogen sulphide ion concentration:

$$2\ SbCl_3 + 6\ [(C_2H_5)_3NH]SH \longrightarrow Sb_2S_3 + 6[(C_2H_5)_3NH]Cl + 3\ H_2S.$$

This reaction may also be regarded as a displacement reaction in which the weak base antimony trisulphide is displaced from its salt by the stronger base triethylammonium hydrogen sulphide.

The fact that solvolysis is preceded by solvation in liquid hydrogen sulphide as well as in other solvents is seen particularly clearly in the dissolution of iodine. This yields brown, conducting solutions, which contain a charge transfer complex of iodine and hydrogen sulphide

$$I_2 \cdot H_2S \longleftrightarrow I_2^- \cdot H_2S^+$$

having a charge transfer band at 290 nm, such as is also found in other solvents in which iodine dissolves to give a brown solution (e.g. pyridine, ammonia, alcohol, ether water). Like many complexes of this type, $I_2 \cdot H_2S$ has an ionic limiting state

$$I_2 \cdot H_2S \longleftrightarrow I_2^- \cdot H_2S^+ \rightleftharpoons I^- + [I \cdot H_2S]^+,$$

which is responsible for the electrical conductivity. The solvation and solvolysis of iodine in hydrogen sulphide are reversible. Careful evaporation of the solvent at low temperatures leaves the deep brown complex $I_2 \cdot H_2S$ behind, and distillation of the solvent at room temperature yields pure iodine. Other solvates of hydrogen sulphide (such as those of beryllium, aluminium, titanium, and tin halides) have also been isolated, but like I_2H_2S these are generally stable only at low temperatures, and their number is very limited.

Hydrogen sulphide also solvolyses organic compounds, frequently in the same way as water. For example, calcium carbide reacts with hydrogen sulphide to form acetylene.

$$CaC_2 + H_2S \longrightarrow C_2H_2 + CaS,$$

though the reaction is very slow. Compounds containing oxygen functions are frequently converted into the corresponding sulphur compounds. For example, thioaldehydes are formed from aldehydes and acyl chlorides give thio acids, which are converted into dithio acids and occasionally, e.g. in the case of dithioacetic acid, even into anhydrides.

$$CH_3COCl + H_2S \longrightarrow CH_3COSH + HCl$$

$$CH_3COSH + H_2S \longrightarrow CH_3CSSH + H_2O$$

$$2\,CH_3CSSH \longrightarrow (CH_3CS)_2S + H_2S.$$

The hydrolysis of esters in water also has parallels in liquid hydrogen sulphide. Methyl thioacetate, for example, is broken down into thioacetic acid and methanethiol:

$$CH_3COSCH_3 + H_2S \rightleftharpoons CH_3COSH + CH_3SH.$$

As in water, the solvolysis of esters in liquid hydrogen sulphide is also an equilibrium reaction. Dithio acids are synthesized by thiohydrolysis of nitriles, e.g.:

$$CH_3CN + H_2S \longrightarrow CH_3C(=S)NH_2$$

$$CH_3C(=S)NH_2 + H_2S \longrightarrow CH_3C(=S)SNH_4$$

$$CH_3C(=S)SNH_4 + HCl \longrightarrow CH_3C(=S)SH + NH_4Cl\downarrow.$$

Thus organic compounds react with liquid hydrogen sulphide if they contain doubly bonded oxygen or triply bonded nitrogen. Other examples of such compounds are cyanamide and dicyanogen. Cyanamide forms thiourea on prolonged standing in a bomb,

$$H_2NCN + H_2S \longrightarrow S = C(NH_2)_2.$$

Dicyanogen dissolves in liquid hydrogen sulphide and reacts via cyanothioformamide to give dithiooxamide:

$$NC - CN + H_2S \longrightarrow NC - C(=S)NH_2$$

$$NC - C(=S)NH_2 + H_2S \longrightarrow H_2N(=S)C - C(=S)NH_2.$$

Further solvolysis of thiourea to trithiocarbonic acid and of dithiooxamide to tetra-thiooxalic acid is not observed. As a weakly acidic compound, hydrogen sulphide also reacts with Grignard compounds:

$$2\,C_2H_5MgBr + H_2S \longrightarrow 2\,C_2H_6 + MgS + MgBr_2.$$

Amphoterism has been investigated for the case of arsenic (III) sulphide. This compound can be solubilized in liquid hydrogen sulphide by triethylammonium hydrogen sulphide or by hydrogen chloride:

$$As_2S_3 + 6\,[(C_2H_5)_3NH](SH) \longrightarrow 2\,[(C_2H_5)_3NH]_3AsS_3 + 3\,H_2S$$

$$As_2S_3 + 6\,HCl \longrightarrow 2\,AsCl_3 + 3\,H_2S.$$

The situation is therefore similar to that of arsenic (III) oxide in water, which is converted into arsenate (III) by alkali and into arsenic (III) chloride by hydrochloric acid.

6.2. Liquid Hydrogen Cyanide

6.2.1. Physico-Chemical Properties of Liquid Hydrogen Cyanide

Hydrogen cyanide, the last of the protonic water-like solvents to be discussed here, is a good solvent for organic and for many inorganic substances, but must be handled with care because of its toxicity. Among its physico-chemical properties (Table 20)

Table 20. Physico-chemical properties of hydrogen cyanide

Molecular weight	27.03
Melting point (°C)	−13.35
Boiling pt. (°C)	+25.0
Density at boiling pt. (g/cm³)	0.681
Molar volume at boiling pt. (cm³)	39.7
Dielectric constant	123 (at 15.6 °C)
Electrical conductance (Ω^{-1})	5×10^{-7} (at 0 °C)
Viscosity (dyn · sec · cm^{-2})	0.00201 (at 20.2 °C)

the high dielectric constant is particularly interesting, and suggests that hydrogen cyanide should be a good solvent for ionic compounds. However, since the molar volume is relatively large (even greater than that of hydrogen sulphide) and the solvation energy consequently low, its solvent power for ionic compounds is poorer than would be expected from the dielectric constant alone.

6.2.2. Solubilities in Liquid Hydrogen Cyanide

The qualitative solubility data for some ionic compounds are given in Table 21.

Table 21. Solubilities of some salts in liquid hydrogen cyanide

Solubility	Solute
Soluble	LiCl, LiBr, LiI, LiSCN, LiClO$_4$, LiNO$_3$, NaBr, NaI, NaNO$_3$, KCl, KBr, KI, KCN, KSCN, KNO$_3$,KHSO$_4$, RbCl, CsCl, FeCl$_3$
Fairly soluble	NaCl, K$_2$SO$_4$, NH$_4$Cl, AgCN, CdI$_2$, HgCl$_2$,As$_2$O$_3$
Insoluble	CaCl$_2$, Ca(NO$_3$)$_2$,BaCl$_2$, Ba(NO$_3$)$_2$, CuSO$_4$, AgCl, AgI, HgI$_2$, PbCl$_2$, PbBr$_2$, PbI$_2$, P$_2$O$_5$.

Many classes of organic substances, such as alcohols, ethers, ketones, acids, acyl halides, and amines (provided that their alkyl groups are not too large) are soluble. Tertiary amines cause the solvent to polymerize, as can be seen from the development of brown colour. Solutions of amines, like those of ionic compounds, are electrically conducting.

6.2.3. Acids, Bases, and Salts

The weak conductivity of pure hydrogen cyanide is due to a slight ionization:

$$2\,HCN \;\rightleftharpoons\; H_2CN^+ + CN^-.$$

According to the solvent theory of acids and bases, therefore, all proton donors are solvo acids in hydrogen cyanide. In practice, the solvo acids are the same as in the aqueous system (e.g. sulphuric acid, trichloroacetic acid, cyanoacetic acid, acetic acid). There are also a few ansolvo acids. For example, antimony (III) chloride in liquid hydrogen cyanide forms a complex chlorocyanoantimonic acid, wich is partly dissociated:

$$SbCl_3 \;+\; 3\,HCN \;\rightleftharpoons\; H_3[SbCl_3(CN)_3] \;\rightleftharpoons\; 3\,H^+ + [SbCl_3(CN)_3]^{3-}$$

Owing to the acidity of the solvent, solvo and ansolvo acids, including those that are known as strong acids in the aqueous system (e.g. hydrogen chloride, nitric acid, sulphuric acid, and trichloroacetic acid), are only slightly dissociated. There appear to be no strong acids in the hydrogen cyanide system.

Solvo bases in hydrogen cyanide are all cyanides. Fairly high degrees of dissociation are found for the ansolvo bases. For example, triethylammonium cyanide is a strong to medium strong electrolyte; it has a degree of dissociation of about 40% even in 0.3 M solution. Though the alkali metal cyanides are solvo bases in hydrogen cyanide, they dissolve only slowly and rapidly turn the solvent brown (polymerization). Only potassium cyanide dissolves rapidly and can therefore be used as a solvo base. Neutralization reactions have been carried out in hydrogen cyanide with the above acids and bases. Some of these reactions were investigated by isolation of the products and others by titration. Nitric acid, which does not oxidize hydrogen cyanide, reacts with potassium cyanide to form potassium nitrate. Neutralization of a solution of the ansolvo base pyridine in hydrogen cyanide with hydrogen chloride gives pyridinium chloride in good yield. Dropwise addition of a solution of potassium cyanide to sulphuric acid leads first to the soluble hydrogen sulphate and, on further addition of potassium cyanide, to the insoluble potassium sulphate.

$$H_2SO_4 + KCN \longrightarrow KHSO_4 + HCN$$

$$KHSO_4 + KCN \longrightarrow K_2SO_4\downarrow + HCN.$$

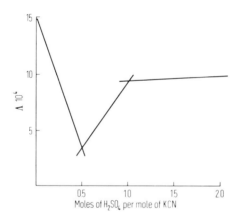

Fig. 7. Conductimetric titration of a solution of potassium cyanide in liquid hydrogen cyanide with sulphuric acid (from G. Jander: Die Chemie in wasserähnlichen Lösungsmitteln. Springer, Berlin, Göttingen, Heidelberg 1949, p. 143, Fig. 26).

On the other hand, when the neutralization of a solution of potassium cyanide by dropwise addition of sulphuric acid in hydrogen cyanide is followed conductimetrically (Fig. 7) the initial decrease in the electrical conductivity shows the formation of the insoluble potassium sulphate K_2SO_4. On further addition of sulphuric acid, the potassium sulphate is redissolved as $KHSO_4$, and the conductivity rises again. The conductivity does not increase appreciably on further addition of acid, since sulphuric acid is only slightly dissociated in liquid hydrogen cyanide. The existence of acid and neutral salts can also be seen from the curves obtained on conductimetric titration of sulphuric acid with triethylamine or propylamine, though these are qualitatively different from the curves obtained with potassium cyanide. An acid salt of the solvo base triethylammonium cyanide with hydrogen chloride is also known. The existence of this compound $[(C_2H_5)_3NH]Cl \cdot HCl$ can also be deduced from a conductimetric titration. In addition to conductimetric and potentiometric titrations, titrations with colour indicators are also possible in hydrogen cyanide. For example, methyl violet is brown-yellow in acid solution, brown-green in neutral solution, and violet-blue in basic solution; in the aqueous system, the same indicator is yellow in acidic solutions and violet in basic solutions.

6.2.4. Solvolysis and Amphoterism

Published data on solvolyses in hydrogen cyanide relate mainly to silver salts. Whereas no sign of solvolysis is found in the case of silver perchlorate (since perchloric acid is a relatively strong acid in hydrogen cyanide), silver nitrate is almost completely solvolysed with formation of insoluble silver cyanide. However, solvolysis can be suppressed by the addition of nitric acid, and 20.02 g of silver nitrate can dissolve without solvolysis in 1 litre of hydrogen cyanide containing 1.92 mole of nitric acid. Silver sulphate is not completely solvolysed; the solvolysis in this case can again be largely suppressed by the addition of sulphuric acid.

An example of the solvolysis of an organic compound in hydrogen cyanide is that of acetyl chloride, which is converted into acetyl cyanide:

$$CH_3COCl + HCN \rightleftharpoons CH_3COCN + HCl.$$

The equilibrium is displaced to the right by the ansolvo base pyridine. This type of reaction is interesting on preparative grounds, since the hydrolysis of the acid cyanide formed by solvolysis leads to keto acids containing one carbon atom more than the original acyl chlorides.

The existence of complex cyanides suggests that amphoterism also occurs in hydrogen cyanide. It is in fact found that e.g. iron(III) cyanide precipitates out from a solution of iron(III) chloride on addition of triethylamine (as triethylammonium cyanide), and redissolves on further addition of amine.

$$FeCl_3 + 3\,[(C_2H_5)_3NH]CN \longrightarrow Fe(CN)_3\downarrow + 3\,[(C_2H_5)_3NH]Cl$$

$$Fe(CN)_3 + 3[(C_2H_5)_3NH]CN \longrightarrow [(C_2H_5)_3NH]_3[Fe(CN)_6].$$

Silver cyanide also exhibits amphoteric behaviour. On the one hand, silver cyanide formed by solvolysis redissolves in hydrogen cyanide containing sulphuric or nitric acid, while on the other silver cyanide precipitated from silver perchlorate with triethylamine is redissolved by an excess of amine:

$$AgClO_4 + [(C_2H_5)_3NH]CN \longrightarrow AgCN \downarrow + [(C_2H_5)_3NH]ClO_4$$

$$AgCN + [(C_2H_5)_3NH]CN \longrightarrow [(C_2H_5)_3NH][Ag(CN)_2].$$

On treatment of a suspension of insoluble mercury cyanide in hydrogen cyanide with potassium cyanide, the mercury salt passes completely into solution at a molar ratio of only $1:1$

$$Hg(CN)_2 + KCN \longrightarrow K[Hg(CN)_3].$$

Further potassium cyanide gives the rather less soluble salt $K_2[Hg(CN)_4]$.
The range of existence of the many known complex cyanides in anhydrous hydrogen cyanide lies in the basic region (high cyanide ion concentration) i.e. on the opposite side of the range of amphoteric compounds from the acid region. The name "peranamphoteric compounds" ($\pi\epsilon\rho\widetilde{\alpha}\nu$ = beyond) has been coined for compounds of this type. This term naturally applies only to complex cyanides that are not, like e.g. trichlorotricyanoantimonic acid, ansolvo acids in liquid hydrogen cyanide. Similarly, the complex hydroxo compounds in water, the complex fluorides in liquid hydrogen fluoride, and the complex amides in liquid ammonia are also peranamphoteric compounds if they are not ansolvo acids in the solvents in question.

6.3. Closing Remarks on the Hydrogen Sulphide and Hydrogen Cyanide

The physico-chemical properties of hydrogen sulphide and hydrogen cyanide and the reactions that take place in them justify the inclusion of these compounds which show weak acidity in comparison with water) among the ionizing solvents. They exhibit (to a smaller extent than water) solvent power for ionic compounds; acids (though only weak) and bases in the sense of the solvent theory are known, as are neutralizations, solvolyses, and amphoterism. Liquid hydrogen sulphide and liquid hydrogen cyanide are also good solvents for substances having covalent bonds, and particularly for organic compounds, as some of the properties of the solvents (e.g. their low degree of association) are similar to those of organic compounds. Experimental difficulties are presented by the inconvenient temperature range in the case of liquid hydrogen sulphide, the tendency toward polymerization in the case of liquid hydrogen cyanide, and the high toxicity of both solvents.

6.4. Bibliography

G. Jander: Die Chemie in wasserähnlichen Lösungsmitteln, Springer, Berlin, Göttingen, Heidelberg 1949, Chapter 4

J. Jander u. G. Türk: Chem. Ber. **95**, 881, 2314 (1962); **98**, 894 (1965)

G. Jander: Die Chemie in wasserähnlichen Lösungsmitteln. Springer, Berlin, Göttingen, Heidelberg 1949, Chapters 5 and 10

7. Acid-Base Concepts and their Usefulness as a Classifying Principle in the Chimistry of Non-Aqueous Ionizing Solvents

7.1. The Solvent Theory

The solvent theory of acids and bases, discussed in Section 1.2, as an extension of the Arrhenius theory of acids and bases in water, readily explains acid-base relationships in protonic solvents, and has recently been used in particular by G. Jander and his co-workers. By concentrating on the dissociation of the solvent and taking the cations and anions of the solvent (i.e. the ions corresponding to protons and to hydroxide ions) as carriers of the acidic and basic functions, it is also possible to consider neutralization, solvolysis, and amphoterism in the various solvents from a common standpoint.

The solvent theory of acids and bases can also be applied to aprotic solvents, provided that they exhibit self-ionization. This is illustrated by the following examples:

$$2\,SO_2 \rightleftharpoons SO^{++} + SO_3^{-}$$

$$2\,I_2 \rightleftharpoons I^{+} + I_3^{-}$$

$$2\,BrF_3 \rightleftharpoons BrF_2^{+} + BrF_4^{-}.$$

As we shall see, however, there is sometimes disagreement regarding the nature and degree of self-ionization of aprotic solvents. Nevertheless, we shall be able to use the solvent theory to a large extent in the discussion of aprotic solvents in the following sections.

7.2. The Electronic Theory; Hard and Soft Acids and Bases

The solvent theory of acids and bases is not the only theory that attempts to explain the behaviour of various substances from the acid-base viewpoint. A concept that is particularly frequently used is the electronic theory of acids and bases due to Lewis. Water can again be used to describe this theory. Whereas the solvent theory generalizes the fact that the proton and the hydroxide ion are the cation and the anion of the solvent, and all compounds that increase the concentration of the solvent cations or anions are defined as acids or bases respectively, the electronic theory generalizes the fact that the proton has a deficiency of electrons while the hydroxide ion has a free pair of electrons. Substances that accept lone pairs of electrons are therefore defined as acids, and substances that can donate lone pairs as bases. This theory, therefore, together with the less well known, but also very comprehensive Usanovich theory[*], is the most general of all acid-base concepts. A few examples are shown in Table 22.

Table 22. Examples of acids and bases in the electronic theory

Acid	Base	Neutralization product
H^+	OH^-	H_2O
$ClCO^+$	Cl^-	$COCl_2$
SO^{2+}	SO_3^-	$2 SO_2$
BCl_3	NR_3	Cl_3BNR_3
Ag^+	$2 NH_3$	$Ag(NH_3)_3^+$

By extension from complex formation reactions, such as are observed with the silver ion, the metal ions in general are nowadays classed as acids. As can be seen from the table, the elements of principal groups 5, 6, and 7 are bases, and Pearson distinguishes between polarizable ("soft") bases and unpolarizable ("hard") bases. Metals that form their most stable complexes with "hard" bases, which contain fluorine, oxygen, or nitrogen, are described according to Pearson as "hard" acids or class a acids, while those that form their most stable complexes with "soft" bases, which contain the heavier elements of these groups, are "soft" acids or class b acids. Thus metals very frequently exhibit the following sequences of complex stability:

for "hard" acids (class a)	for "soft" acids (class b)
$N \gg P > As > Sb > Bi$	$N \ll P > As > Sb > Bi$
$O \gg S > Se > Te$	$O \ll S \approx Se \approx Te$
$F \gg Cl > Br > I$	$F < Cl < Br \ll I$

On comparison of the stabilities, not just within one group of the periodic system, but in general, the following sequences are found:

---

[*] According to Usanovich, all positive coordinately unsaturated atomic groupings are acids and all negative coordinately saturated atomic groupings are bases. According to this theory practically any ionic reaction is an acid-base reaction.

for "hard" acids (class a)	for "soft" acids (class b)
approximately the reverse of that for class b	$C \approx S > I > Br \approx Cl > N > O > F.$

There are "hard" acids that form complexes in water only with oxide or fluoride ligands. "Soft" acids, on the other hand, also form complexes with carbon monoxide, olefins, aromatic hydrocarbons and similar compounds.
Some "hard" and "soft" acids are listed in Table 23.

Table 23. Examples of "hard" and "soft" acids

"Hard" acids (class a)	"Soft" acids (class b)
H^+, Li^+, Na^+, K^+, Be^{2+}, Mg^{2+},	Cu^+, Ag^+, Au^+, Tl^+, Hg^+, Cs^+,
Ca^{2+}, Sr^{2+}, Sn^{2+}, Al^{3+}, Sc^{3+},	Pd^{2+}, Cd^{2+}, Pt^{2+}, Hg^{2+}, CH_3Hg^+,
Ga^{3+}, In^{3+}, La^{3+}, Cr^{3+}, Co^{3+},	Tl^{3+}, $Tl(CH_3)_3$, BH_3, RS^+, RSe^+,
Fe^{3+}, As^{3+}, Ir^{3+}, Si^{4+}, Ti^{4+},	RTe^+, I^+, Br^+, HO^+, RO^+, I_2,
Zr^{4+}, Th^{4+}, Pu^{4+}, VO^{2+}, $UO_2{}^{2+}$,	Br_2, ICN, Trinitrobenzene
$(CH_3)_2Sn^{2+}$, $Be(CH_3)_2$, BF_3, BCl_3,	Chloranil, quinones, zero-valent
$B(OR)_3$, $Al(CH_3)_3$, $RPO_2{}^+$, $ROPO_2{}^+$,	metal atoms
$ROS_2{}^+$, $ROSO_2{}^+$, SO_3, I^{7+}, I^{5+},	
Cl^{7+}, R_3C^+, RCO^+, CO_2, NC^+, HX	
(Molecules that form hydrogen bonds)	

Limiting cases

Fe^{2+}, Co^{2+}, Ni^{2+}, Cu^{2+}, Zn^{2+}, Pb^{2+}, $B(CH_3)_3$, SO_2, NO^+.

The metals of class a ("hard" acids) therefore include all small unpolarizable metal cations with high oxidation numbers (frequently with an inert gas configuration); class b acids ("soft" acids) are larger and have lower oxidation numbers. "Hard" acids prefer "hard" bases, and "soft" acids "soft" bases. The type of bonding between the acid and the base is of minor importance; it is predominantly ionic in the case of "hard" acids, and predominantly covalent in the case of "soft" acids. As can be seen from Table 22, the classification of an element is not constant, but depends on its oxidation state. The acid character of an element is also influenced by groups bound to it. Groups that give up negative charges make the acid "softer", since they decrease the oxidation number by reduction. Groups that readily donate electrons, e.g. the hydride or the sulphide ion, are readily polarizable, and are therefore "soft" bases.

That the reaction of a metal ion with a ligand to form a complex is in fact comparable to a neutralization in accordance with the solvent theory can be seen e.g. from the reaction of aluminium ions ("hard" acids) with a chelating agent ("hard" base). The concentration of aluminium ions is found to change in the same way as the pH in acidimetry or alkalimetry.

The acid-base concept of the electronic theory is the most general known. At the same time, however, it is the weakest. By defining a wide range of different substances as acids or bases in the same way, it loses its value as a classifying principle. The electronic theory is even less suitable for use as a classifying principle for chemistry in ionizing solvents, since, in order to be as generally valid as possible, it does not consider the solvent. This is a grave disadvantage, in view of the enormous influence of the ionizing solvents on the reactions that occur in them. Smith tried to overcome this disadvantage by means of a modified electronic theory in which the acid-base definition is based on the solvent (a substance is an acid if it acts as an acceptor of electron pairs with respect to the molecule or an ion of the solvent, while it is a base if it can act as an electron donor). Though this theory takes the solvent as the reference substance for the acid-base defintion, it is less selective than the solvent theory with respect to the choice of acids and bases. It can however be used as a classifying principle for solvating solvents that exhibit practically no self-ionization (cf. Section 7.4).

7.3. The Ionotropic Theory

To be able to understand the ionotropic theory of acids and bases proposed by Gutmann and Lindquist, it is necessary to recall Brønsted's theory of the acid-base function (proton-transfer theory). This theory is again derived from the acid-base concepts for water, and defines acids as substances that can initiate a proton transfer by giving up protons, and bases as substances that can terminate a proton transfer by accepting protons. Some examples are shown in Table 24.

Table 24. Acids and bases in the proton-transfer theory

Acid 1	+	Base 2	$\rightleftharpoons$ Acid 2	+	Base 1
H_3O^+	+	OH^-	$\rightleftharpoons H_2O$	+	H_2O
NH_4^+	+	OH^-	$\rightleftharpoons NH_3$	+	H_2O
HCl	+	NH_3	$\rightleftharpoons NH_4^+$	+	Cl^-
H_2O	+	HS^-	$\rightleftharpoons H_2S$	+	OH^-
HSO_4^-	+	H_2O	$\rightleftharpoons H_3O^+$	+	SO_4^{2-}

The proton-transfer theory can be applied to solvents in which proton transfer plays a dominant role. It can therefore be used as a classifying principle for protonic ionizing solvents such as liquid ammonia or liquid hydrogen fluoride. It naturally fails, however, in solvents such as sulphur dioxide or bromine trifluoride, which contain no protons.

In the ionotropic theory, proton transfer (prototropy) is merely a special case. The general acid-base definition in this theory is based on ion transfer (ionotropy). It can therefore also be applied to aprotic ionizing solvents and their typical ion transfers (e.g. in self-ionization). Solvents are described as cationotropic or anionotropic, according to whether cations or anions are transferred in the solvent. The prototropic solvent systems of the proton-transfer theory are thus cationotropic solvents in the ionotropic theory. The acid-base definitions of the proton-transfer theory are valid for cationotropic systems if the word "cation" is inserted instead of "proton". In anionotropic solvent systems, on the other hand, ion donor and acceptor functions are reversed. Thus acids and bases in their most general form are defined as follows: An acid is a cation donor in a cationotropic solvent system or an anion acceptor in an anionotropic system, cation and anion being the respective classes of ions that undergo the ion transfer. A base is a cation acceptor in a cationotropic system or an anion donor in an anionotropic solvent system.

Some cationotropic and anionotropic solvents are listed in Table 25. The anionotropic solvents can be further subdivided into fluoridotropic, chloridotropic, oxidotropic, etc. systems, whereas the only cationotropic solvents known are those that exchange protons.

Table 25. Cationotropic and anionotropic solvents

Cationotropic solvents
(acids: proton donors, bases: proton acceptors)

Base 2	+ Acid 1	$\rightleftharpoons$ Acid 2	+ Base 1
H_2O	+ H_2O	$\rightleftharpoons$ H_3O^+	+ OH^-
NH_3	+ NH_3	$\rightleftharpoons$ NH_4^+	+ NH_2^-
N_2H_4	+ N_2H_4	$\rightleftharpoons$ $N_2H_5^+$	+ $N_2H_3^-$
HF	+ HF	$\rightleftharpoons$ H_2F^+	+ F^-
H_2SO_4	+ H_2SO_4	$\rightleftharpoons$ $H_3SO_4^+$	+ HSO_4^-
CH_3COOH	+ CH_3COOH	$\rightleftharpoons$ $CH_3COOH_2^+$	+ CH_3COO^-
C_2H_5OH	+ C_2H_5OH	$\rightleftharpoons$ $C_2H_5OH_2^+$	+ $C_2H_5O^-$
CH_3CONH_2	+ CH_3CONH_2	$\rightleftharpoons$ $CH_3CONH_3^+$	+ CH_3CONH^-

Anionotropic solvents
(Fluoridotropic solvents (acids: fluoride ion acceptors, bases: fluoride ion donors)

Base 2	+ Acid 1	$\rightleftharpoons$ Acid 2	+ Base 1
BrF_3	+ BrF_3	$\rightleftharpoons$ BrF_2^+	+ BrF_4^-
IF_5	+ IF_5	$\rightleftharpoons$ IF_4^+	+ IF_6^-
$3\,KF$	+ AlF_3	$\rightleftharpoons$ $3\,K^+$	+ AlF_6^{3-}

chloridotropic solvents
(acids: chloride ion acceptors, bases: chloride ion donors)

Base 2	+ Acid 1	$\rightleftharpoons$ Acid 2	+ Base 1
$AsCl_3$	+ $AsCl_3$	$\rightleftharpoons$ $AsCl_2^+$	+ $AsCl_4^-$
ICl	+ ICl	$\rightleftharpoons$ I^+	+ ICl_2^-
$POCl_3$	+ $POCl_3$	$\rightleftharpoons$ $POCl_2^+$	+ $POCl_4^-$
$SOCl_2$	+ $SOCl_2$	$\rightleftharpoons$ $SOCl^+$	+ $SOCl_3^-$
$COCl_2$	+ $COCl_2$	$\rightleftharpoons$ $COCl^+$	+ $COCl_3^-$

Bromidotropic solvents (acids: bromide ion acceptors, bases: bromide ion donors)

Base 2	+ Acid 1	$\rightleftharpoons$ Acid 2	+ Base 1
$HgBr_2$	+ $HgBr_2$	$\rightleftharpoons HgBr^+$	+ $HgBr_3^-$
IBr	+ IBr	$\rightleftharpoons I^+$	+ IBr_2^-

Oxidotropic solvents (acids: oxygen ion acceptors, bases: oxygen ion donors)

Base 2	+ Acid 1	$\rightleftharpoons$ Acid 2	+ Base 1
SO_2	+ SO_2	$\rightleftharpoons SO^{2+}$	+ SO_3^{2-}
Oxide melts			
CaO	+ CO_2	$\rightleftharpoons Ca^{2+}$	+ CO_3^{2-}
Na_2O	+ SiO_2	$\rightleftharpoons 2\,Na^+$	+ SiO_3^{2-}

Just as acidity in prototropic solvents can be expressed by the pH value, so anion activites can be indicated by corresponding quantities such as pO, pF, or pCl values. These quantities may be taken as a direct measure of the basicity of a solution, since the anion concentration, i.e. the basicity, increases with decreasing p(anion) value. Provided that we consider only the self-ionization of the solvent, the ionotropic theory is very similar to the solvent theory of acids and bases, though it always treats acids and bases as acidbase pairs. However, these pairs need not be identical with the species existing in equilibrium as a result of the self-ionization of the solvent. For example, tin (IV) chloride in the chloridotropic arsenic(III) chloride is an acid (chloride acceptor) both in the ionotropic theory and in the solvent theory, whereas the arsenic (III) chloride, unlike in the solvent theory, is a base (chloride donor) in the ionotropic theory:

Base 2	+ Acid 1	$\rightleftharpoons$ Acid 2	+ Base 1
$2\,AsCl_3$	+ $SnCl_4^-$	$\rightleftharpoons 2\,AsCl_2^+$	+ $SnCl_6^{2-}$

The solvent theory recognizes only the acid $(AsCl_2^+)_2(SnCl_6^{2-})$, bases being e.g. $(M_4N^+)(AsCl_4^-)$, i.e. compounds that can split off the solvent cation $AsCl_2^+$ and the solvent anion $AsCl_4^-$ respectively.

The definitions of acids and bases in cationotropic or anionotropic solvents are also compatible with the electronic theory of acids and bases. As in the solvent theory, however, they have the advantage of closer connection with the solvent, which very often has a strong effect on the manifestation of acidic or basic functions.

7.4. The Coordination Model

In the solvent theory (and in the ionotropic theory), the selfionization of the solvent forms the basis of the acid-base definition, and also of the acid-base act (e.g. solvoly or formation of ansolvo acids and bases). As we have seen, the solvents discussed so far react in accordance with the solvent theory, and give no reason to abandon it. There are however solvents in which, though a slight self-ionization is observed, this can no longer serve as a foundation for the formation and reactions of acids and bases. These solvents are borderline cases between water-like solvents and those that are no longer water-like (cf. Section 1.1). They are mainly organic solvents, and particularly those containing nitrogen, but they also include e.g. phosphorus oxide trichloride and sulphur dioxide. In these solvents the acid-base act is initiated not by self-ionization but by solvation (coordination) of the dissolved compounds (coordination model proposed by Meek and Drago). This can be illustrated for a solution of iron (III) chloride in phosphorus oxide trichloride. According to the solvent theory acid-base definition and the acid-base act would depend on the self-ionization of the solvent as follows:

$$POCl_3 \rightleftharpoons POCl_2^+ + Cl^-.$$

Added iron (III) chloride would thus form tetrachloroferrate (III) ions

$$FeCl_3 + POCl_3 \rightleftharpoons POCl_2^+ + FeCl_4^-,$$

and would therefore behave as an ansolvo acid (increasing the solvent cation concentration). The solvent should thus give up chloride ions to the iron (III) chloride. According to the coordination model, the acid-base act depends on the solvation of iron (III) chloride

$$2\,FeCl_3 + 6\,POCl_3 \rightleftharpoons 2\,FeCl_3(OPCl_3)_3 \rightleftharpoons FeCl_2(OPCl_3)_4^+ + FeCl_4^- + 2\,POCl_3$$

i.e. a halogen transfer takes place between solvates of the type $FeCl_3(OPCl_3)_3$. The chloride ion taken up by one molecule of iron (III) chloride thus comes from a second molecule of iron (III) chloride. A subsequent equilibrium

$$FeCl_2(OPCl_3)_4^+ + POCl_3 \rightleftharpoons FeCl_3 + POCl_2^+ + 4\,POCl_3$$

would lead to the same products as in the solvent theory; in a nonchloridotropic solvent this equilibrium would not be established and the solvated dichloroiron ion would act as the acidic species. The coordination model is experimentally confirmed by the fact the $FeCl_4^-$ ion is also formed on dissolution of iron (III) chloride in triethyl phosphate $PO(OC_2H_5)_3$, in which chloride ions cannot result from self-ionization of the solvent. It is also known that e.g. in the solvates $AlCl_3 \cdot OPCl_3$ and $SbCl_5 \cdot OPCl_3$ the phosphorus oxide trichloride is coordinated via the oxygen, and not via chlorine.

It can be seen that the self-ionization of the solvent is of no importance in some cases in coordinating solvents, and it is therefore not taken into account in the coordination model. The only important factor in this model is the ability to form solvates. In the absence of self-ionization, the effect of the solvent on the acid-base definition that characterizes the solvent theory is also absent, and the effect on the acid-base act is greatly weakened. The modified electronic theory offers a suitable acid-base definition for solvents that react in accordance with the coordination model. Other Lewis acids besides iron (III) chloride also behave as described above toward coordinating solvents (Lewis bases) (cf. Section 7.2).

7.5. Bibliography

G. Jander: Die Chemie in wasserähnlichen Lösungsmitteln. Springer, Berlin, Göttingen, Heidelberg 1949, Chapter 10

W.F. Luder and S. Zuffanti: The Electronic Theory of Acids and Bases. Wiley, New York 1946; M. Usanowitsch, J. allg. Chem. (U.S.S.R.) **9**, 182 (1939); R.G. Pearson, J. Amer. Chem. Soc. **85**, 3533 (1963); Nachr. Chem. Techn. **13**, 237 (1965); G.B.L. Smith, Chem. Rev. **23**, 165 (1938)

V. Gutmann u. I. Lindqvist, Z. physik. Chem. **203**, 250 (1954)

D.W. Meek: Lewis Acid-Base Interactions in Polar Non-Aqueous Solvents, in J.J. Lagowski: The Chemistry of Non-Aqueous Solvents, Academic Press, New York, London 1966, Vol. I, Chapter 1; R.S. Drago u. K.F. Purcell: Coordinating Solvents, in T.C. Waddington: Non-Aqueous Solvent Systems. Academic Press, London, New York 1965, Chapter 5; J.E. Huheey, J. inorg. nucl. Chem. 24,1011 (1962).

8. Liquid Sulphur Dioxide

8.1. Physico-Chemical Properties of Liquid Sulphur Dioxide

Liquid sulphur dioxide was examined as an ionizing solvent by Walden et al. as early as the turn of the century. It has steadily gained in interest as a readily liquefiable, selective solvent in particular for covalent compounds. Care is necessary in handling it, however, owing to its toxicity. Its physico-chemical properties are shown in Table 26.

Table 26. Physico-chemical properties of sulphur dioxide

Melting point, (°C)	−75.46
Boiling point, (°C)	−10.02
Heat of fusion (kcal/mol)	1.9691
Heat of vaporization (kcal/mol)	5.96
Vapour pressure (atm)	1.530 (at 0 °C)
	3.228 (at 20 °C)
Viscosity (Poise)	0.004285 (at 0 °C)
Dielectric constant	15.6 (at 0 °C)
	12.35 (at 22 °C)
Dipole moment (Debye)	1.62
S-O bond length (Å)	1.43
O-S-O bond angle (degrees)	119.5
Density (g/cm^3)	1.46 (at −10 °C)
Specific conductivity ($\Omega^{-1}cm^{-1}$)	$3 \times 10^{-8} - 4 \times 10^{-8}$

The electronic structure of the sulphur dioxide molecule is best represented by the resonance hybrids (a) to (c). It is clear from hybrid (c) that sulfur dioxide can act as an electron acceptor.

8.2. Solubilities in Liquid Sulphur Dioxide; Solvation

Ionic substances exhibit only slight solubility in liquid sulphur dioxide, owing to the relatively low dielectric constant of the solvent. Only salts with relatively large cat-

ions [e.g. $(CH_3)_4N^+$] or anions (e.g. SCN^-, I^-), i.e. salts having low lattice energies are readily soluble. The solubilities of the halides therefore normally increase toward the iodides, as was seen earlier in the case of liquid ammonia (Section 2.2.1). For example, sodium chloride is insoluble, whereas sodium bromide has a solubility of 1.36 and sodium iodide one of 1000 mmole/1000 g of SO_2. The same order of solubilities, i.e. the opposite of that found in water, is also observed for the silver halides, though even silver iodide has a solubility of only 0.68 mmole/1000 g of SO_2. The thiocyanates and cyanides are relatively soluble (KSCN 502.0, KCN 2.62, NH_4SCN 6160.0, AgSCN 0.845, AgCN 1.42 mmole/1000 g of SO_2). Lithium fluoride has a surprisingly high solubility (23.0 mmole/1000 g of SO_2). Alkali metal sulphites are not very soluble and give concentrations of only 10^{-3} to 10^{-2} M. Sulphites with larger cations, such as the alkylammonium sulphites, on the other hand, are readily soluble. These are converted into pyrosulphites, $M_2S_2O_5 = M_2SO_3 \cdot SO_2$ by solvation. Sulphates, sulphides, oxides, and hydroxides are normally insoluble.
As is expected from the dielectric constant of the solvent, covalent substances are more soluble. Bromine, iodine monochloride, thionyl chloride, arsenic trichloride, boron trichloride, carbon disulphide, phosphorus trichloride and phosphorus oxide chlorides are completely miscible with liquid sulphur dioxide. Carbon, silicon, tin, and lead tetrachlorides are also soluble above a critical mixing temperature. Thionyl bromide, thionyl cyanate, and other thionyl compounds known only in liquid sulphur dioxide are also readily soluble.
The solubilities of many compounds can be strongly influenced by water. For exampl anhydrous cobalt (II) thiocyanate is insoluble in liquid sulphur dioxide, whereas blue $[Co(H_2O)_2](SCN)_2$ dissolves in the presence of small quantities of moisture. This colour reaction can be used for the detection of small quantities of water in liquid sulphur dioxide. Water itself has a solubility of 2.3 g/100 g of SO_2 at 22 °C.
The selective solvent properties of liquid sulphur dioxide for organic substances is very important and is used in industry. For example, the solubility of olefins and aromatic hydrocarbons in sulphur dioxide is used in the petroleum industry to separate these compounds from saturated hydrocarbons, which become less soluble with increasing molecular weight (Edeleanu process). It is possible in this way to remove the olefinec and aromatic components of a petroleum fraction, which improve a petrol, from the saturated components, which lower the octane number. However, sulphur dioxide also dissolves the organic sulphides and disulphides, which are undesirable in petrol, so that it is necessary to start with low sulphur petroleum fractions to obtain good petrols having high octane numbers. The components that are insoluble in sulphur dioxide can be used in diesel fuels. Other groups of organic substances that are readily soluble in sulphur dioxide are amines, esters, alcohols, phenols, and carboxylic acids.

It should be mentioned that liquid sulphur dioxide can form solvates, those of iodides, thiocyanates, and tetramethylammonium salts being particularly well known. Some examples of formulae are $NaI \cdot 2 SO_2$, $NaI \cdot 4 SO_2$, $KI \cdot 4 SO_2$, $KBr \cdot 4 SO_2$, $KSCN \cdot 0.5 SO_2$, $KSCN \cdot SO_2$, $LiI \cdot 2 SO_2$, $CsI \cdot 4 SO_2$, $CaI_2 \cdot 4 SO_2$, $SrI_2 \cdot 2 SO_2$, $SrI_2 \cdot 4 SO_2 \cdot BaI_2 \cdot 2 SO_2$, $BaI_2 \cdot 4 SO_2$, $(CH_3)_4NBr \cdot SO_2$, $(CH_3)_4NBr \cdot 2 SO_2$, $(CH_3)_4NCl \cdot SO_2$, $(CH_3)_4NCl \cdot 2 SO_2$, $[(CH_3)_4N]_2SO_4 \cdot 3 SO_2$. In the solvates of the alkali metal halides, the sulphur dioxide is bound to the anion, a charge transfer interaction occuring between the anion and the sulphur dioxide, i.e. the negative charge of the anion is partly transferred to the sulphur dioxide.

Solvation also occurs with covalent compounds. Examples of known adducts are BF_3SO_2, $SbF_5 \cdot SO_2$, $SnBr_4 \cdot SO_2$, $AlCl_3 \cdot SO_2$, and $H_2O \cdot SO_2$. Special mention should be made of the adduct formed at temperatures below -50 °C by iodine and liquid sulphur dioxide, which may be regarded as a charge transfer complex with iodine as the donor and sulphur dioxide as the acceptor. Interesting solvates, many of which are coloured, are formed by the amines. For example, triethylamine gives a red adduct having the composition $(C_2H_5)_3N \cdot SO_2$. It is not yet certain whether this adduct, which may also be regarded as a charge transfer complex, dissociates further as follows:

$$ 2\ R_3N\ +\ 2\ SO_2\ \rightleftharpoons\ 2\ R_3\overset{\delta+}{N}\rightarrow\overset{\ddot{O}:}{\underset{\ddot{O}:}{\overset{\delta-}{S}}}\ \rightleftharpoons\ [(R_3N)_2SO]^{2+}\ +\ SO_3{}^{2-} $$

p-Phenylenediamine gives the adduct $p\text{-}C_6H_4(NH_2)_2 \cdot 2 SO_2$, and triethylamine oxide gives $(C_2H_5)_3NO \cdot SO_2$. Sulphur dioxide also enters into charge transfer interactions with alcohols and phenols, as well as with a number of aromatic hydrocarbons. The charge transfer complexes can be recognized by the appearance of a new absorption band in the UV spectrum, the excitation energy of this band corresponding to the energy of the complete charge transfer from the donor to the acceptor.

8.3. Electrochemical Studies

Solutions of alkali metal, ammonium, and tetraalkylammonium chlorides, bromides, iodides, thiocyanates, and tetrafluoroborates in liquid sulphur dioxide conduct electricity relatively well, though the conductivity is lower than that of corresponding solutions in water or liquid ammonia. This is as expected from the difference in the dielectric constants of the three solvents ($H_2O > NH_3 > SO_2$). Monoalkylammonium, dialkylammonium and trialkylammonium salts are very poor conductors in liquid sulphur dioxide, as are the metal sulphites and thionyl compounds, the proble-

matic definition of which as bases and acids will be discussed later (Section 8.4). Con
ducting solutions are also obtained with compounds that have covalent structures
but can also form ionic limiting states. These include the halogens bromine and iodine, interhalogen compounds (IBr, ICl, ICl_3), halogen compounds of the group 5
elements (e.g. PBr_3, PBr_5, $AsBr_3$, $SbCl_5$), amines, various ketones, carbinols and free
radicals [e.g. $(C_6H_5)_3COH$], triaryl halides [e.g. $(C_6H_5)_3CCl$], acyl chlorides and
bromides [e.g. $(CH_3)_2CHCOBr$], and some polyethers.

Since liquid sulphur dioxide has a low dielectric constant, as the dilution of the
above 1,1 electrolytes increases, the molecular conductivity Λ_v passes through a
minimum at about 10^{-1} mole/litre and then increases on further dilution to the
value of the limiting conductivity Λ_∞. The dilution at which this value is reached
is generally very much higher in sulphur dioxide than in water. An unusual feature
in this case is the solubility maximum observed before the minimum at medium
concentrations. It must be assumed that ion pairs and higher aggregates are present
in this concentration range.

In agreement with the conductivity measurements, ebullioscopic and cryoscopic
determinations indicate that non-electrolytes are monomolecular in solution, whereas salts are more or less strongly dissociated in dilute solutions and associated in
more concentrated solutions. The latter is particularly true of the thiocyanates.

The conductivity limits of the cations, which correspond to the ion mobilities, increase in the following order:

$$Li^+ < (C_2H_5)_4N^+ < (CH_3)_4N^+ < K^+ < Na^+.$$

Increasing ion mobility should correspond to decreasing ionic size, but it can be seen
that this is not the case. The deviation from the expected order, as in water, is explained by differences in the solvation of the ions. The mobilities of anions increase
with decreasing size:

$$ClO_4^- < I^- < Br^- < Cl^-.$$

The calculated degrees of dissociation of 1,1 salts show that for a given anion, the
dissociation increases according to the nature of the cation in the following order:

$$Li^+ < Na^+ < K^+ < (CH_3)_4N^+ < (C_2H_5)_4N^+.$$

The dissociation of a salt, and hence its solubility, thus increases with increasing
ionic size (Section 8.2). For a given cation, the dissociation depends on the nature
of the anion.

$$BF_4^- < ClO_4^- < Cl^- < Br^- < I^-.$$

The deviation from the expected order is again attributed to differences in solvation. Electrolyses have also been carried out in liquid sulphur dioxide. It is interesting to note that the hydronium ion H_3O^+ has been detected on electrolysis of a mixture of water and hydrogen bromide in liquid sulphur dioxide. Neither component alone gives a conducting solution in this solvent; when they are mixed, however, the solution becomes conducting and water and hydrogen are liberated at the cathode and bromine at the anode. On electrolysis of a solution of triphenylmethyl chloride, the triphenylmethyl radical is obtained at the cathode.

8.4. Acids and Bases

The very slight conductivity of liquid sulphur dioxide (cf. Section 8.1) is explained by a weak self-ionization as follows:

$$2 SO_2 \rightleftharpoons SO^{2+} + SO_3^{2-}$$
$$SO_3^{2-} + SO_2 \rightleftharpoons S_2O_5^{2-}.$$

According to the solvent theory of acids and bases, therefore, sulphites or disulphites may be regarded as bases, and compounds that yield SO^{2+} cation as acids. This "sulphito" system of acids and bases, which has been largely used by G. Jander since 1936, will be retained in the following discussion for systematic reasons, though recent investigations have shown that its value is to a large extent only formal. It is now known from exchange reactions with radioactive oxygen (^{18}O), sulphur (^{35}S), and chlorine that acid-base reactions in the sulphito system can be explained without the self-ionization of sulphur dioxide, i.e. without the occurrence of SO^{2+} cations. Thus sulphur dioxide molecules exchange their oxygen atoms even in the gas phase or in cyclohexane; this exchange can be explained by the formation of cyclic associates of the type

$$O = S \underset{O}{\overset{O}{\diamond}} S = O$$

The dissociation of thionyl chloride in liquid sulphur dioxide

$$SOCl_2 \rightleftharpoons SO^{2+} + 2 Cl^-$$

which should occur according to the solvent theory, does not take place, since thionyl chloride does not exchange either sulphur or oxygen with the solvent. It dissociates instead in accordance with the equation

$$SOCl_2 \rightleftharpoons SOCl^+ + Cl^-,$$

as is shown by the rapid halogen exchange between thionyl chloride and chloride ions. On the other hand, thionyl chloride and thionyl bromide in liquid sulphur dioxide exchange radioactively labelled sulphur with each other. This could result either from direct halogen exchange or via a cyclic complex

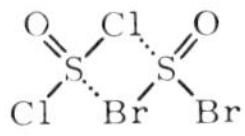

which then dissociates into molecules of the type SOClBr.

It is interesting in this connection to note that sulphur exchange between thionyl compounds and the solvent can be catalysed by the addition of various halogen compounds. For example, the addition of tetramethylammonium bromide to a solution of thionyl bromide leads to sulphur exchange, the rate of which depends only on the concentration of the tetramethylammonium bromide, and which can be explained by the following scheme:

$$SOBr_2 \rightleftharpoons SOBr^+ + Br^- \qquad\qquad \text{fast}$$

$$^*SO_2 + Br^- \rightleftharpoons {}^*SO_2Br^- \qquad\qquad \text{fast}$$

$$^*SO_2Br^- + SO_2 \rightleftharpoons {}^*SOBr^+ + SO_3^{2-} \quad \text{slow}$$

The sulphur exchange observed with thionyl chloride in the presence of tetramethyl-ammonium chloride proceeds rather differently:

$$^*SO_2 + Cl^- \rightleftharpoons {}^*SO_2Cl^- \qquad\qquad\qquad \text{fast}$$

$$^*SO_2Cl^- + SOCl_2 \rightleftharpoons {}^*SOCl_2 + SO_2Cl^- \qquad \text{slow}$$

Nonionic compounds such as triethylamine, acetone, antimony pentachloride and aluminium chloride also have a catalytic action, the mechanism of which is not yet fully known.

As is shown by the exchange experiments, the occurrence of free SO^{2+} cations in sulphur dioxide solutions of thionyl compounds is only an assumption based on the solvent theory of acids and bases (cf. also the very low conductivity of these solutions. Section 8.1). Acids defined by the self-ionization of the solvent are thus unknown in liquid sulphur dioxide. The reaction of thionyl chloride with the sulphite ion is therefore not a neutralization in accordance with the solvent theory

$$SO^{2+} + 2\,Cl^- + SO_3^{2-} \rightleftharpoons 2\,SO_2 + 2\,Cl^-,$$

in which undissociated solvent is formed from solvent cations and anions, but can be explained by the electronic theory as a nucleophilic attack of the Lewis base SO_3^{2-} on the Lewis acid $SOCl_2$ with expulsion of two chloride ions.

$$SO_3^{2-} + SOCl_2 \rightleftharpoons \left\{SO_3 \cdot SOCl_2^{2-}\right\} \longrightarrow \left\{SO_3 \cdot SOCl^-\right\} + Cl^-$$
$$\longrightarrow \left\{SO_3 \cdot SO\right\} + 2\,Cl^- \longrightarrow 2\,SO_2 + 2\,Cl^-$$

and presumably also occurs in this way in other inert solvents. It is possible that as in the case of phosphorus oxide trichloride, the process is influenced by solvation (cf. Section 7.4).

Despite these limitations, we shall retain the solvent concept of acids and bases in the following discussion since it has proved satisfactory as a classifying principle for chemistry in liquid sulphur dioxide. The most important "acid" in liquid sulphur dioxide is thionyl chloride, which can be prepared as follows:

$$PCl_5 + SO_2 \longrightarrow POCl_3 + SOCl_2.$$

Thionyl chloride is sensitive to moisture, and decomposes above 80 °C into disulphur dichloride, chlorine, and sulphur dioxide. Thionyl bromide is prepared by the action of dry hydrogen bromide on thionyl chloride

$$SOCl_2 + 2\,HBr \longrightarrow SOBr_2 + 2\,HCl$$

or by the reaction of phosphorus(V) bromide with liquid sulphur dioxide:

$$PBr_5 + SO_2 \longrightarrow POBr_3 + SOBr_2.$$

Other thionyl compounds such as thionyl thiocyanate, thionyl acetate, and thionyl phenylacetate can be obtained by the reaction of a suitable salt with thionyl chloride in liquid sulphur dioxide:

$$2\,NH_4SCN + SOCl_2 \longrightarrow 2\,NH_4Cl\downarrow + SO(SCN)_2$$

$$2\,CH_3COOAg + SOCl_2 \longrightarrow 2\,AgCl\downarrow + SO(CH_3COO)_2$$

$$2\,C_6H_5CH_2COONH_4 + SOCl_2 \longrightarrow 2\,NH_4Cl\downarrow + SO(C_6H_5CH_2COO)_2.$$

Another method for the preparation of thionyl compounds is the reaction of anhydrides with the solvent sulphur dioxide:

$$(CH_3CO)_2O + SO_2 \longrightarrow SO(CH_3COO)_2$$

$$(CH_2ClCO)_2O + SO_2 \longrightarrow SO(CH_2ClCOO)_2.$$

Antimony trichloride reacts with thionyl chloride in liquid sulphur dioxide to give the "acid" thionyl hexachloroantimonate (III):

$$3\,SOCl_2 + 2\,SbCl_3 \rightleftharpoons (SO)_3[SbCl_6]_2.$$

Some of these thionyl compounds are known only in solution. Like carbonic acid and sulphurous acid in water, they decompose when the solvent is distilled off. Thionyl iodide is unknown.

Some examples of "neutralizations" are given in the following equations:

$$Na_2SO_3 + SOCl_2 \longrightarrow 2\,SO_2 + 2\,NaCl\downarrow$$

$$Cs_2SO_3 + SOCl_2 \longrightarrow 2\,SO_2 + 2\,CsCl\downarrow$$

$$K_2S_2O_5 + SO(SCN)_2 \longrightarrow 3\,SO_2 + KSCN$$

$$[(CH_3)_4N]_2SO_3 + SOBr_2 \longrightarrow 2\,SO_2 + 2\,[(CH_3)_4N]Br.$$

These "neutralizations" can be followed conductimetrically as well as by preparative-analytical investigations. For example, in the reaction of potassium pyrosulphite with thionyl thiocyanate, the conductivity curve shows a sharp change in direction at a molar ratio of 1 : 1.

8.5. Solvolyses

Solvolysis reactions in liquid sulphur dioxide are less common, slower, and often more complex in nature than in protonic solvents. For example, alkali metal bromides and iodides undergo slow solvolysis, which appears in the case of potassium bromide to consist of the following reactions:

$$8\,KBr + 8\,SO_2 \longrightarrow 4\,K_2SO_3 + 4\,SOBr_2$$
$$4\,SOBr_2 \longrightarrow 2\,SO_2 + S_2Br_2 + 3\,Br_2$$
$$\underline{4\,K_2SO_3 + 2\,Br_2 \longrightarrow 2\,K_2SO_4\downarrow + 4\,KBr + 2\,SO_2}$$
$$4\,KBr + 4\,SO_2 \longrightarrow 2\,K_2SO_4\downarrow + S_2Br_2 + Br_2.$$

A similar reaction course is assumed for potassium iodide, except that the expected disulphur diiodide is unstable. The overall equation in this case is:

$$4\,KI + 4\,SO_2 \longrightarrow 2\,K_2SO_4\!\downarrow + 2\,S + 2\,I_2.$$

It has recently been found that solutions of the bromides and iodides in sulphur dioxide are stable in the absence of oxygen, and that the instability in particular of the iodides is due to the oxygen dissolved in sulphur dioxide. Solvolysis reactions have been observed in particular with halides of the elements of groups 5 and 6 and of the elements of the group 5 and 6 triads. Whereas the halides of the group 4 elements are not solvolysed, phosphorus(V) chloride is solvolysed in liquid sulphur dioxide with formation of phosphorus oxide trichloride and thionyl chloride even at $-50\,°C$ (see also Section 8.4).

$$PCl_5 + SO_2 \longrightarrow POCl_3 + SOCl_2.$$

Phosphorus pentabromide reacts similarly:

$$PBr_5 + SO_2 \longrightarrow POBr_3 + SOBr_2.$$

Antimony trichloride and antimony pentachloride are not solvolysed. Solvolyses have also been observed with niobium(V), uranium(V) and tungsten(VI) chorides:

$$NbCl_5 + SO_2 \xrightarrow{70\,°C} NbOCl_3 + SOCl_2$$

$$WCl_6 + SO_2 \xrightarrow{70\,°C} WOCl_4 + SOCl_2\,(\text{red needles on cooling})$$

$$2\,UCl_5 \xrightarrow{90\,°C} UCl_6 + UCl_4$$

$$UCl_6 + 2\,SO_2 \longrightarrow UO_2Cl_2\!\downarrow + 2\,SOCl_2.$$

The sparingly soluble anhydrous acetates also undergo slow solvolysis in liquid sulphur dioxide. Sulphites and disulphites are formed, and the reaction can be used e.g. in the case of ammonium acetate for the preparation of anhydrous sulphate-free ammonium sulphite:

$$2\,CH_3COONH_4 + 2\,SO_2 \longrightarrow (NH_4)_2SO_3 + SO(CH_3COO)_2.$$

The thionyl acetate formed as an intermediate is converted into acetic anhydride and sulphur dioxide when the solvent is distilled off.

The solvolysis of an organometallic compound has also been described. Diethyl zinc reacts rapidly with liquid sulphur dioxide even at the temperature of an ether-carbon dioxide bath in accordance with the equation

$$Zn(C_2H_5)_2 + SO_2 \longrightarrow ZnO + O{=}S(C_2H_5)_2.$$

8.6. Amphoterism

Addition of the "base" tetramethylammonium sulphite to a solution of aluminium chloride in liquid sulphur dioxide leads to precipitation of aluminium sulphite, which is dissolved by an excess of the "base" to give the peranamphoteric sulphitoaluminate

$$2\,AlCl_3 + 3\,(Me_4N)_2SO_3 \longrightarrow Al_2(SO_3)_3{\downarrow} + 6\,Me_4NCl$$

$$Al_2(SO_3)_3 + 3\,(Me_4N)_2SO_3 \longrightarrow 2\,(Me_4N)_3Al(SO_3)_3.$$

When the "acid" thionyl chloride is added to such a solution, aluminium sulphate is reprecipitated:

$$2\,(Me_4N)_3Al(SO_3)_3 + 3\,SOCl_2 \longrightarrow Al_2(SO_3)_3{\downarrow} + 6\,Me_4NCl + 3\,SO_2$$

Gallium trichloride behaves similarly. Tin sulphite can be precipitated from a solution of tin(IV) chloride with the "base" tetramethylammonium sulphite, but reacts with an excess of the precipitant to give soluble, peranamphoteric tetrasulphitostannate(IV), $(Me_4N)_4Sn(SO_3)_4$. Tin itself is also soluble in sulphur dioxide/tetramethylammonium sulphite:

$$Sn + 2\,(Me_4N)_2S_2O_5 + SO_2 \longrightarrow (Me_4N)_2Sn(SO_3)_3 + (Me_4N)_2S_2O_3.$$

This reaction is analogous to the dissolution of the metal in strong alkali solutions. Amphoteric behaviour has also been observed for non-metal compounds. For example phosphorus trichloride reacts with tetramethylammonium sulphite to form phosphorus trioxide, which reacts with further sulphite to give a soluble peranamphoteric compound:

$$2\,PCl_3 + 3\,(Me_4N)_2SO_3 \longrightarrow P_2O_3{\downarrow} + 3\,SO_2 + 6\,Me_4NCl$$

$$P_2O_3 + (Me_4N)_2S_2O_5 \longrightarrow 2\,(Me_4N)PO_2SO_2.$$

Antimony(III) and antimony(V) oxides also exhibit amphoterism.

8.7. Complex Formation Reactions

Using sulphur dioxide as the solvent, Seel prepared a series of complexes that cannot be obtained from aqueous solution. Thus the reaction of nitrosyl chloride with antimony pentachloride yields nitrosyl hexachloroantimonate:

$$NOCl + SbCl_5 \longrightarrow NO[SbCl_6].$$

The compound $NO_2[SbCl_6]$ can be prepared from nitryl chloride and antimony pentachloride. Acetyl chloride reacts with antimony(V) chloride as follows:

$$CH_3COCl + SbCl_5 \longrightarrow [CH_3CO][SbCl_6].$$

Reactions of this type can be followed conductimetrically. Antimoy pentachloride reacts with $SOCl_2$ to give the adduct $SOCl_2 \cdot SbCl_5$, which does not exist in solution either as $SO[SbCl_6]_2$ as was formerly assumed, or as $SOCl[SbCl_6]$. The structure is probably similar to that of the adduct $SeOCl_2 \cdot SbCl_5$, for which coordinate bonding of the oxygen to the pentachloride has been detected. Complexes of antimony also occur on reaction of antimony pentachloride with potassium iodide:

$$6\ KI + 3\ SbCl_5 \longrightarrow 3\ I_2 + 6\ KCl + 3\ SbCl_3$$

$$6\ KCl + 2\ SbCl_3 \longrightarrow 2\ K_3[SbCl_6].$$

The reaction of potassium chloride with antimony pentachloride yields the complex $K[SbCl_6]$ which is also formed from potassium hexachloroantimonate(III) and excess antimony(V) chloride:

$$2\ K_3[SbCl_6] + 6\ SbCl_5 \longrightarrow 6\ K[SbCl_6] + 2\ SbCl_3,$$

giving the overall equation

$$2\ KI + 3\ SbCl_5 \longrightarrow I_2 + 2\ K[SbCl_6] + SbCl_3$$

for the reaction of potassium iodide with excess antimony pentachloride. This is confirmed by conductimetric measurements. The resulting hexachloroantimonate(V) can enter into double decompositions, as is shown by the following equations:

$$[CH_3CO][SbCl_6] + KCl \longrightarrow CH_3COCl + K[SbCl_6]$$

$$[CH_3CO][SbCl_6] + NOCl \longrightarrow CH_3COCl + NO[SbCl_6]$$

$$NO[SbCl_6] + (CH_3)_4NClO_4 \longrightarrow (CH_3)_4N[SbCl_6] + NOClO_4\downarrow.$$

Complex formation also occurs with aluminium and boron halides. Thus acetyl chloride and boron trifluoride give $CH_3CO[BF_4]$ while benzoyl chloride and aluminium chloride give $C_6H_5CO[AlCl_4]$. Similarly, the solubility of cadmium iodide or mercury iodide in liquid sulphur dioxide is increased by the addition of potassium iodide, this behaviour indicating complex formation.

Some iodine complexes formed by the action of iodine on iodides are also known in liquid sulphur dioxide, e.g.:

$$RbI + I_2 \longrightarrow RbI_3$$

$$KI + I_2 \longrightarrow KI_3.$$

8.8. Organic Reactions in Liquid Sulphur Dioxide

Owing to its incombustibility, its solvent power for covalent compounds, its low reactivity, and its relatively convenient liquid range, sulphur dioxide is also useful as a solvent for reactions in organic chemistry. For example, Friedel-Crafts syntheses proceed in good yields and with satisfactory rates in liquid sulphur dioxide.

$$C_6H_6 + tert.-C_5H_{11}Cl \xrightarrow{AlCl_3} tert.-C_5H_{11}C_6H_5 + HCl$$

$$C_6H_6 + C_6H_5COCl \xrightarrow{AlCl_3} C_6H_5COC_6H_5 + HCl.$$

Liquid sulphur dioxide can be used as a solvent for the bromination of organic compounds. Examples are the brominations of styrene and of phenol:

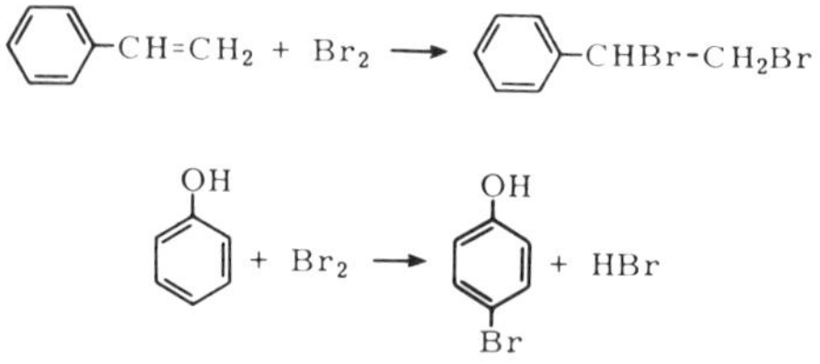

Solutions of bromine in sulphur dioxide can be readily converted into anhydrous hydrogen bromide solutions by the addition of the calculated quantity of water:

$$Br_2 + 2H_2O + SO_2 \longrightarrow 2HBr + H_2SO_4.$$

A solution produced in this way reacts with styrene to give 1-bromo-2-phenylethane in 70% yield.

Liquid sulphur dioxide is also an excellent solvent for the sulphonation of aromatic compounds with sulphur trioxide or chlorosulphonic acid. The yields are usually very good, and the products are not contaminated with inorganic salts:

$$C_6H_6 + SO_3 \longrightarrow C_6H_5SO_3H$$

$$C_6H_6 + ClSO_3H \longrightarrow C_6H_5SO_3H + HCl.$$

When the mixture obtained on reaction of cyclohexanecarboxylic acid in liquid sulphur dioxide with fuming nitric and sulphuric acid at −20 °C, is slowly warmed and poured onto ice, it yields ε-caprolactam, the starting material for the preparation of Perlon, in 60% yield.

8.9. Closing Remarks on the Sulphur Dioxide System

Liquid sulphur dioxide is an aprotic solvent, which has a medium dielectric constant and is a relatively good solvent for covalent compounds as well as for some ionic substances. Many compounds having relatively high lattice energies are even more soluble than would be expected from the dielectric constant of sulphur dioxide. This solvent also has a good ionizing power. For example, as in liquid hydrogen chloride (dielectric constant = 9.3), triphenylmethyl chloride is largely dissociated in liquid sulphur dioxide (dielectric constant = 15.3), whereas the same compound is practically undissociated in many other solvents (e.g. nitrobenzene, dielectric constant = 24.5). The dissociation and solubility of this halide, and presumably of all other halides, are probably due to the formation of charge transfer complexes, the halides acting as electron donors and the solvent molecules as acceptors.

Though acid-base relationships in liquid sulphur dioxide can be formally described in accordance with the solvent theory as reactions of thionyl compounds ("sulphito" bases), exchange reactions with isotopes show that free SO^{2+} ions do not occur in the "neutralizations". True acids in the sense of the solvent theory of acids and bases are therefore unknown in liquid sulphur dioxide.

8.10. Bibliography

G. Jander: Die Chemie in wasserähnlichen Lösungsmitteln. Springer, Berlin, Göttingen, Heidelberg 1949, Chapter 8

L.F. Audrieth u. J. Kleinberg: Non-Aqueous Solvents, Wiley, New York 1953, Chapter 11

T.C. Waddington: Liquid Sulphur Dioxide, in T.C. Waddington: Non-Aqueous Solvent Systems, Academic Press, London, New York 1965, Chapter 6

H.H. Sisler: Chemistry in Non-Aqueous Solvents, Reinhold, New York 1961

J.C.L. Defize: On the Edeleanu Process. D.B. Centen's Uitgevers Mataschappij N.V. Amsterdam 1938

J. Jander u. G. Türk: Angew. Chem. **75**, 792 (1963)

N. Tokura, T. Kawahara u. T. Sato: Bull. Chem. Soc. Japan **38**, 849 (1965); Angew. Chem. **78**, 341 (1966)

9. Halogens and Interhalogen Compounds

9.1. Physico-Chemical Properties of the Halogens and Interhalogen Compounds

The chemistry of the halogens and interhalogen compounds as solvents is not very well known. One reason for this is the high reactivity of these compounds; a solute cannot only enter into the usual reactions (dissociation, solvation, solvolysis, etc.), but may also be decomposed by the solvent. Another reason is the difficulty of finding a suitable material for apparatus. For example, bromine trifluoride, which reacts explosively with water and organic compounds, can be handled in quartz vessels, but even quartz in the compact state is slowly attacked. The iodine chlorides even dissolve gold and platinum. Nevertheless, some of these solvents are very interesting from the point of view of preparative possibilities.
Some of their physico-chemical properties are given in Table 27.

9.2. The Solvent Properties of the Halogens

Little is known as yet about chemistry in liquid chlorine. Sodium, potassium, copper (II), cadmium, aluminium, lead(II), and zirconium(IV) chlorides have been found to be insoluble in this solvent. Carbon(IV), silicon(IV), titanium(IV), lead(IV), and arsenic(III) chlorides and phosphorus oxide trichloride are readily soluble. Carbon tetrachloride forms solvates of the type $CCl_4 \cdot nCl_2$ where n=0.5, 1, 2, 3, and 4. Similar solvates are known for chloroform, methylene chloride, and hydrogen chloride. There is as yet no evidence of the existence of the ion Cl_3^- in liquid chlorine or of the formation of complexes e.g. with antinony pentachloride. Solubilities in bromine are very similar to those in chlorine. For example, boron(III) bromide, carbon(IV) chloride and bromide, and silicon(IV), tin(IV), titanium(IV), arsenic(III) and antimony(III) bromides are readily soluble, as are bromides having large cations such as tetrabutylammonium bromide, which gives viscous, conducting solutions. Alkali metal bromides dissolve only to a very small degree. The relatively high solubility of a caesium bromide (large cation) which is equal to 19.3 g/100 g of bromine at 25 °C, is of industrial importance, since caesium can be obtained in a purity of 99.7% by extraction of a mixture of rubidium and caesium bromides with bromine. The rubidium content can be further decreased by recrystallization from water. The pure caesium bromide obtained in this way is finely used to obtain pure caesium, which plays an important part in ionic drives and thermionic energy converters.

Table 27. Physico-chemical properties of some halogens and interhalogen compounds

	I_2	IF_5	ICl	IBr	Br_2	BrF_3	Cl_2
Melting point (°C)	113.6	9.6	27.2(α) 13.9(β)	42	-7	8.8	-101
Boiling pt (°C)	184.4	98	100	-116	59	127.6	-34
Density (g/cm³)	3.92 (133 °C)	3.33 (9.6 °C)	3.13 (45 °C)	3.76 (42 °C)	3.12 (20 °C)	2.84 (8.8 °C)	1.57 (-35 °C)
Dielectric constant	11.1 (118 °C)	36.2 (35 °C)	—	—	3.1 (20 °C)	—	2.0 (-35 °C)
Specific conductivity ($\Omega^{-1} \cdot cm^{-1}$)	5.2×10^{-5} (114 °C)	1.6×10^{-5} (9.6 °C)	4.4×10^{-3} (27.2 °C)	4.0×10^{-4} (42 °C)	1.1×10^{-9} (25 °C)	8.1×10^{-3} (8.8 °C)	7×10^{-8}
Viscosity (cPoise)	1.98 (116 °C)	2.19 (25 °C)	4.19 (28.4 °C)	—	1.03 (16 °C)	2.22 (25 °C)	4.9 (-34 °C)
Trouton constant (cal · mol⁻¹ · deg.⁻¹)	23.1	27.2	26.7	—	22.0	25	20.4

Bromine reacts with a number of organic bases such as pyridine, quinoline, and acetamide to form adducts, which give electrically conducting solutions in bromine and probably consist of the ions $[(Base)_n Br^+]$ and Br_3^-.

Liquid iodine is the most extensively investigated of these solvents. Iodine itself has special properties. Thus intermolecular interactions occurring in the solid state persist in the liquid state, as is shown by the value of the Trouton constant. The dielectric constant increases with rising temperature (11.1 at 118 °C, 13.0 at 168 °C). The electric conductivity of molten iodine is greater than that of the other halogens, but decreases with rising temperature. This is explained by the assumption that the conductivity is partly metallic and partly ionic.

Solubility data in liquid iodine are mainly qualitative. The alkali metal iodides and ammonium iodide are soluble. Thallium(I); mercury(II), phosphorus(III), arsenic(III) antimony(III), bismuth(III), tin(IV), and iron iodides also dissolve without chemical change. Of the elements, sulphur, selenium and tellurium dissolve without chemical change, whereas the metals aluminium, tin, arsenic, antimony, bismuth, and iron dissolve with formation of the iodides. Organic compounds such as iodoform, tetra-methylammonium iodide, p-dibromobenzene, naphthalene, diphenyl, and benzoic acid are also reported to be soluble in molten iodine. A number of metals, such as copper, silver, lead, nickel, and platinum, are insoluble, since they are protected from further attack by a solid film of iodide. A platinum-iridium alloy containing 15 % of iridium appears to be most stable to iodine, and is therefore used as an electrode material in this solvent.

Some iodides form solvates (polyiodides) having the following formulae: $RbI \cdot I_2$, $CsI \cdot I_2$, CsI_4, and $ZrI_4 \cdot I_2$. Most of the solvates are unstable above the melting point of iodine. The solutions of iodides are very complex in nature, since dissociation into the ions is accompanied by the formation of polymeric species, For example, cryoscopic measurements show that lithium iodide has a degree of polymerization of 3–5, while the degree of polymerization of caesium iodide is only about 1.4. At the same time, the solutions exhibit electrical conductivities that can only be explained by the formation of more complex ions, e.g. I_3^-. It is not clear why the specific conductivity of sodium iodide is lower than that of lithium, potassium, or rubidium iodide. The conductivity of pure molten iodine is probably due to slight self-ionization:

$$2 I_2 \rightleftharpoons I^+ + I_3^-$$

The concentration of the I^+ ions is very low, and the ionic product $[I^+] \cdot [I^-]$ is reported to be 10^{-42}. More complex cations e.g. of the type I_3^+, are probably also present in molten iodine. The I^+ cation occurs in rather higher concentrations in the blue solutions of iodine in oleum or iodine pentafluoride, where it can be identified by its paramagnetism. From the self-ionization of iodine, it is possible according to the solvent theory to define acids and bases as substances that can provide the iodine cation and the iodide ion respectively. Bases in liquid iodine therefore include iodides and triiodides, while the acids include compounds such as iodine chloride, iodine bromide, and iodine cyanide. It is not known whether the iodine-pyridine complexes of the type $[I (Py)]NO_3$ and $[I (Py)_2]NO_3$, which also contain positive iodine, are acids in the iodine system.

The reaction of potassium iodide with iodine bromide in iodine is a neutralization:

$$KI_3 + IBr \longrightarrow KBr + 2 I_2.$$

However, this reaction probably does not proceed via free iodine cations, since the iodine halides are only weak electrolytes in molten iodine. It is possible that the applicability of the solvent theory to iodine is subject to the same reservations as in sulphur dioxide (cf. Sections 7.4 and 8.4). Other "neutralizations" are represented by the following equations:

$$HgI_2 + 2 IBr \longrightarrow HgBr_2 + 2 I_2$$

$$BiI_3 + 3 ICl \longrightarrow BiCl_3 + 3 I_2.$$

Reactions of this type can be followed conductimetrically. The relatively high conductivity e.g. of a solution of rubidium iodide is rapidly reduced by the addition of iodine bromide until at a molar ratio of 1 : 1 it reaches a value that remains practically constant on further addition of iodine bromide.

The inclusion of molten iodine among the ionizing solvents is also justified by the fact that solvolysis and amphoterism are observed in this system. For example, the cyanides of sodium, potassium, silver, and mercury are solvolysed in iodine to the corresponding iodides:

$$KCN + I_2 \longrightarrow KI + ICN.$$

The reaction proceeds particularly readily because iodine cyanide is practically insoluble, and sublimes off. Examples of amphoteric behaviour in liquid iodine are provided by lead(II) and bismuth(III) iodides, which are formed from the corresponding chlorides with potassium iodide and give complexes having the compositions $K[PbI_3]$ and $K_3[BiI_6]$ in an excess of potassium iodide. Mercury iodide forms $K_2[HgI_4]$.

9.3. The Solvent Properties of the Interhalogen Compounds

Some of the properties of iodine monochloride are typical of ionizing solvents. For example, solutions of potassium, ammonium, rubidium, aluminium, and phosphorus(V) chlorides and of potassium bromide and iodide have higher electrical conductivities than the pure solvent. Solutions of silicon(IV) and titanium(IV) chlorides are practically non-conducting. The conductivity of pure iodine chloride is explained by the self-ionization

$$2\,ICl \rightleftharpoons I^+ + ICl_2^-$$

though the existence of free iodine cations, as in liquid iodine, is doubtful. The cation I_2Cl^+ is probably present. The anion ICl_2^- has been detected by X-ray methods, e.g. in the solid compound $[PCl_4][ICl_2]$. The reaction of antimony penta-chloride with the solvent can probably be formulated as follows:

$$2\,ICl + SbCl_5 \rightleftharpoons I_2Cl[SbCl_6].$$

Antimony pentachloride is thus an ansolvo acid according to the solvent theory. The potassium chloride solvate $KCl \cdot ICl = K[ICl_2]$ is a base according to this theory. The neutralization of the two substances can therefore be formulated as follows:

$$K[ICl_2] + I_2Cl[SbCl_6] \longrightarrow K[SbCl_6] + 3\,ICl.$$

The titration of antimony(V) chloride with potassium chloride in iodine chloride can be followed conductimetrically, and gives a conductivity curve with a sharp salient point at a molar ratio of 1 : 1.

The situation in iodine monobromide is very similar to that in iodine chloride. According to the solvent theory, compounds that release the ion IBr_2^- should be regarded as bases. Thus the alkali metal bromides give polyhalides having the formula $M[IBr_2]$ on dissolution in iodine bromide; phosphorus pentabromide gives the compound $IPBr_6$, which probably exists as $PBr_4[IBr_2]$. Neutralizations with the ansolvo acid tin(IV) bromide:

$$2\,RbIBr_2 + SnBr_4 \rightleftharpoons Rb_2SnBr_6 + 2\,IBr$$

can be followed conductimetrically.

Little is known about iodine trichloride and chlorine trifluoride as solvents. Iodine trichloride is dimeric and it is probable from its structure and conductivity that it contains ICl_2^+ cations and ICl_4^- anions: these also occur on dissolution of acids (e.g. aluminium chloride) and bases (e.g. potassium chloride) in this system. Chlorine trifluoride dissociates slightly in accordance with

$$2\,ClF_3 \rightleftharpoons ClF_2^+ + ClF_4^-.$$

Compounds containing the anion ClF_4^- (e.g. rubidium tetrafluorochlorate $Rb[ClF_4]$) and the cation ClF_2^+ (e.g. difluorochlorine tetrafluoroborate $ClF_2[BF_4]$) can be obtained; neutralizations between these compounds however have not yet been attempted.

The most interesting solvent of this group is bromine trifluoride since it has gained considerable importance for the preparation of complex fluorides. Bromine trifluoride is a very strong fluorinating agent, so that solubility studies are confined to inorganic fluorides. The alkali metal fluorides, silver, barium, gold(III), boron, titanium(IV), vanadium(V), niobium(V), phosphorus(V), arsenic(V), and antimony(V) fluorides are soluble. From the self-ionization of bromine trifluoride.

$$2\,BrF_3 \rightleftharpoons BrF_2^+ + BrF_4^-$$

the compounds listed from gold fluoride onwards may according to the solvent theory be regarded as acids (e.g. $BrF_2[SbF_6]$), and the alkali metal fluorides, silver fluoride, and barium fluoride as bases (e.g. $Ag[BrF_4]$). Neutralizations can be followed conductimetrically

$$BrF_2[SbF_6] + Ag[BrF_4] \longrightarrow Ag[SbF_6] + 2\,BrF_3.$$

Further examples of this type of reaction are provided by the following equations:

$$(BrF_2)_2[SnF_6] + 2\,K[BrF_4] \rightleftharpoons K_2[SnF_6] + 4\,BrF_3$$

$$VF_5 + Ag[BrF_4] \longrightarrow Ag[VF_6] + BrF_3$$

$$BrF_2[TaF_6] + LiF \longrightarrow Li[TaF_6] + BrF_3.$$

Dissolution of gold and silver together in bromine trifluoride leads directly to silver tetrafluoroaurate:

$$Ag + Au \xrightarrow{BrF_3} Ag[BrF_4] + BrF_2[AuF_4] \longrightarrow Ag[AuF_4] + 2\,BrF_3.$$

The preparation of other complex fluorides is illustrated by the following equations:

$$NOCl + SnF_4 \xrightarrow{BrF_3} (NO)_2[SnF_6]$$

$$N_2O_4 + Sb_2O_3 \xrightarrow{BrF_3} (NO_2)[SbF_6]$$

$$Ru + KCl \xrightarrow{BrF_3} K[RuF_6].$$

Bromine trifluoride also stabilizes some higher oxidation states that are otherwise difficult to obtain: For example palladium(IV) fluoride, which has not so far been isolated, forms the salt $K_2[PdF_6]$. Chromium(V) forms the complex $K[CrOF_4]$. Iodine pentafluoride must also be regarded as an ionizing solvent; it dissociates to a small degree, presumably in accordance with

$$2\,IF_5 \rightleftharpoons IF_4^+ + IF_6^-.$$

Potassium chloride and antimony pentafluoride considerably increase the conductivity of iodine pentachloride:

$$KF + IF_5 \longrightarrow K[IF_6]$$

$$SbF_5 + IF_5 \longrightarrow IF_4[SbF_6]$$

$$K[IF_6] + IF_4[SbF_6] \longrightarrow K[SbF_6] + 2\,IF_5.$$

Dissolution of iodine or iodine tetroxide in iodine pentafluoride leads to blue solutions containing the paramagnetic iodine cation:

$$6\,IF_5 + 2\,I_2 \rightleftharpoons 5\,I^+ + 5\,IF_6^-.$$

Very little is known about iodine heptafluoride as an ionizing solvent. Its self-ionization probably proceeds in accordance with the equation:

$$2\,IF_7 \rightleftharpoons IF_6^+ + IF_8^-.$$

The cation IF_6^+ is probably present in the addition compound $AsF_5 \cdot IF_7 (=IF_6[AsF_6])$. No compounds containing the IF_8^- anion are known as yet.

9.4. Bibliography

H. Spandau u. V. Gutmann: Angew. Chem. **64,** 93 (1952)

L.F. Audrieth u. J. Kleinberg: Non-Aqueous Solvents. Wiley, New York 1953, Chapter 13

A.G. Sharpe: The Halogens and Interhalogens as Solvents, in T.C. Waddington: Non-Aqueous Solvent Systems. Academic Press, London, New York 1965, Chapter 7

V.A. Stenger: Angew. Chem. **78,** 315 (1966)

G. Jander: Die Chemie in wasserähnlichen Lösungsmitteln. Springer, Berlin, Göttingen, Heidelberg 1949, Chapter 7

H. Krebs: Z. Elektrochem., Ber. Bunsenges. physik. Chem. **61,** 933 (1957).

10. Survey of Other Ionizing Solvents

The foregoing sections have dealt with a selection of protonic and aprotic solvents having ionizing properties that appear characteristic to the authors. However, this book would be incomplete without reference to other solvents, some organic and others inorganic, that round off the picture of ionizing water-like solvents available at present. This brief survey of these solvents will begin with the halides of group 5 elements, and the related melts of half salts (salts having only partial ionic character) will be represented by a particularly typical example, mercury(II) bromide. These will be followed by some oxide halides and dinitrogen tetroxide, and the discussion will end with the organic ionizing solvents in common use.

10.1. Inorganic Solvents

10.1.1. Physico-Chemical Properties

The principal physico-chemical properties of the solvents to be discussed here are listed in Table 28.

Both the solvent theory of acids and bases (Section 7.1) and the ionotropic theory (Section 7.3) can be used with some reservations (cf. Section 7.4) as classifying principles for the halides and for the oxide halides discussed later. For example, arsenic(III) chloride is assumed to dissociate (though only slightly) as follows:

$$2 AsCl_3 \rightleftharpoons AsCl_2^+ + AsCl_4^-$$

According to the solvent theory, therefore, acids in arsenic(III) chloride are substances that increase the concentration of $AsCl_2^+$ ions, while bases are substances that increase the concentration of $AsCl_4^-$ ions. According to the ionotropic theory, arsenic(III) chloride is a chloridotropic (anionotropic) solvent, and acids and bases are defined as chloride ion acceptors and donors respectively. The $AsCl_3$ molecule can therefore act as an acid by accepting chloride ions to form the ion $AsCl_4^-$ or as a base by giving up chloride ions to form the ion $AsCl_2^+$. Solid compounds containing the $AsCl_4^-$ anion are known, examples being the potassium and tetramethyl-ammonium salts $K[AsCl_4]$ and $Me_4N[AsCl_4] \cdot 2AsCl_3$; however these readily lose arsenic(III) chloride above 160 °C.

Table 28. Physico-chemical properties of some inorganic ionizing solvents

	Melting point (°C)	Boiling point (°C)	Dielectric constant	Specific conductivity ($\Omega^{-1} \cdot cm^{-1}$)	Viscosity (cPoise)	Density (g/cm³)	Molecular weight
AsF_3	−6	63	5.7 (<−6 °C)	2.4×10^{-5} (25 °C)	—	2.67 (0 °C)	131.91
$AsCl_3$	−13	130	12.6 (17 °C)	1.4×10^{-7}	1.225 (20 °C)	2.16 (20 °C)	181.28
$AsBr_3$	35	220	8.8 (35 °C)	1.6×10^{-7} (35 °C)	5.41 (35 °C) 4.44 (40 °C)	3.33 (50 °C)	314.66
$SbCl_3$	73	221	33.0 (75 °C)	0.85×10^{-6} (95 °C)	3.3 (95 °C)	2.44 (178 °C)	228.13
$SbBr_3$	97	280	20.9 (100 °C)	0.9×10^{-5}–1×10^{-5} (100 °C)	6.81 (100 °C)	4.15 (23 °C)	361.51
$BiCl_3$	232	447	—	—	32.0 (260 °C)	4.75 (25 °C)	315.37
$HgBr_2$	238	320.3	9.84 (solid)	1.45×10^{-4} (242 °C)	3.698 (240 °C)	5.109 (242 °C)	360.44
$COCl_2$	−128	8.2	4.34 (22 °C)	7×10^{-9} (25 °C)	—	1.435 (0 °C)	98.92
$NOCl$	−61.5	−6	22.5 (−27,5 °C)	2.88×10^{-6} (−20 °C)	0.586 (−27 °C)	1.59 (−6 °C)	65.47
$POCl_3$	1	108	13.9 (22 °C)	2×10^{-8} (20 °C)	1.15 (25 °C)	1.71 (0 °C)	153.35
$SeOCl_2$	10.9	176	46.2 (20 °C)	2.0×10^{-5} (25 °C)	—	2.424 (22 °C)	165.87
N_2O_4	−12.3	21.3	2.42 (18 °C)	2.36×10^{-13} (17 °C)	—	1.49 (0 °C)	92.02

10.1.2. Halides

Metals and metal oxides, sulphates, and nitrates are only very sparingly soluble in arsenic(III) chloride. Alkali metal and ammonium chlorides dissolve more readily, while the chlorides of aluminium, tin(IV), vanadium(IV), and iron(III) or compounds such as $(CH_3)_4N[SbCl_6]$ are readily soluble.

Solutions of these substances have high conductivities since they contain bases (chloride ion donors) or ansolvo acids (chloride ion acceptors):

$$(CH_3)_4NCl + AsCl_3 \rightleftharpoons (CH_3)_4N^+ + AsCl_4^-$$

$$SnCl_4 + AsCl_3 \rightleftharpoons AsCl_2^+ + SnCl_5^- .$$

The formation of tetramethylammonium hexachloroantimonate in arsenic(III) chloride by titration of antimony(V) chloride with tetramethylammonium chloride can be detected conductimetrically; and is a neutralization according to the solvent theory:

$$(CH_3)_4N[AsCl_4] + AsCl_2[SbCl_6] \longrightarrow (CH_3)_4N[SbCl_6] + 2 AsCl_3 .$$

Similarly, tin(IV) chloride reacts with tetramethylammonium chloride in accordance with the following equation:

$$(CH_3)_4N[AsCl_4] + SnCl_4 \rightleftharpoons (CH_3)_4N[SnCl_5] + AsCl_3 .$$

The $(CH_3)_4N[SnCl_5]$ is converted into $[(CH_3)_4N]_2[SnCl_6]$ on further addition of chloride. Titanium(IV) chloride does not appear to react with arsenic(III) chloride, and its solutions are practically non-conducting.

Since no conductivity minimum is found on conductimetric titration of phosphorus(V) chloride in $AsCl_3$ with tetramethylammonium chloride, it was concluded that phosphorus(V) chloride is not an ansolvo acid (chloride ion acceptor, formation of $AsCl_2$ $[PCl_6]$) but a base (chloride ion donor, formation of PCl_4 $[AsCl_4]$). A solid having the composition $2 PCl_5 . 5 AsCl_3$ can be isolated from a solution of phosphorus(V) chloride in arsenic(III) chloride but the structure of this product has not been established in detail.

Conductimetric titration of a solution of tellurion(IV) chloride in arsenic(III) chloride with tetramethylammonium chloride gives a conductivity curve having two salient points (at molar ratios of 1 : 1 and 1 : 2). The compounds $(CH_3)_4NCl \cdot TeCl_4 \cdot AsCl_3$ (which can probably be regarded as the "acid" salt $[(CH_3)_4N](AsCl_2)[TeCl_6])$ and $2 (CH_3)_4NCl \cdot TeCl_4$ (which may be regarded as $[(CH_3)_4N]_2[TeCl_6])$ are formed. Phosphorus pentachloride gives a still more complex titration curve, which indicates

the presence of the compounds $(PCl_4)\,(AsCl_3)_3\,[TeCl_6]_2$, $(PCl_4)_2[TeCl_6]$. The first
two compounds readily lose arsenic(III) chloride to give the compounds $PCl_5 \cdot 2TeC$
and $PCl_5 \cdot TeCl_4$. However, the reaction of tellurium (IV) chloride with tin (IV)
chloride in arsenic (III) chloride indicates that tellurium (IV) chloride acts as a base
(chloride ion donor) and tin (IV) chloride as an acid (chloride ion acceptor):

$$(AsCl_2)_2[SnCl_6] + TeCl_3[AsCl_4] \rightleftharpoons (TeCl_3)\,(AsCl_2)\,[SnCl_6] + 2\,AsCl_3.$$

Similarly, antimony(V) chloride gives a compound $TeCl_3\,[SbCl_6]$. Tellurium(IV)
chloride thus evidently exhibits amphoteric properties in arsenic(III) chloride.
There is therefore only indirect (stoichiometric and conductimetric) evidence of the
existence of the ions $AsCl_2{}^+$ and $AsCl_4{}^-$ in solution. As in solid compounds (see
above), the $AsCl_4{}^-$ ion is evidently not very stable in solution, as is shown inter alia
by the relatively high mobility of the chloride ion in arsenic(III) chloride, which can
be explained by a chloriue ion transfer from tetrachloroarsenate ions to arsenic(III)
chloride molecules.
To explain the slight conductivity of arsenic(III) fluoride and arsenic(III) bromide,
it has been assumed that these compounds dissociate in the same way as arsenic(III)
chloride. However, the existence of the ions $AsF_2{}^+$ and $AsF_4{}^-$ in arsenic(III) fluoride
is uncertain. Arsenic (III) fluoride is a good fluorinating agent, but has so far found
little use as a solvent. More information is available about arsenic(III) bromide which
dissolves e.g. quaternary ammonium bromides and boron, aluminium, gallium, tin(IV)
titanium(IV), phosphorus(III), phosphorus(V), and antimony(III) bromides, whereas
the bromides of the alkali metals, alkaline earth metals, and divalent transition metals
are insoluble. Many organic compounds such as hydrocarbons, alcohols, ketones, ester
and amines are soluble. Tertiary amines behave as ansolvo bases, and give conducting
solutions in which equilibria of the type

$$R_3N + AsBr_3 \rightleftharpoons R_3NAsBr_2{}^+ + Br^-$$

probably occur.
The existence of the $AsBr_2{}^+$ ion is shown by the solvolysis that takes place on reaction of silver perchlorate with arsenic(III) bromide at 50 °C.

$$AsBr_3 + AgClO_4 \longrightarrow AsBr_2[ClO_4] + AgBr \downarrow.$$

Though the compound $AsBr_2\,[ClO_4]$ cannot be isolated, its presence is shown by
the high conductivity of the solution. The $AsBr_2{}^+$ cation is also formed on dissolution
of ansolvo acids (bromide ion acceptors) such as aluminium, gallium, tin(IV) and
bismuth(III) bromides in arsenic(III) bromides. The solutions of these compounds can

be titrated with bases (bromide ion donors), such as tetraethylammonium bromide:

$$Et_4NBr + AsBr_2[AlBr_4] \longrightarrow Et_4N[AlBr_4] + AsBr_3.$$

Neutralization reactions of this type can be followed conductimetrically and potentiometrically. The advantage of arsenic(III) bromide as a solvent is that it is possible in this way to prepare complex bromides that cannot be obtained in any other way, such as $K[AlBr_4]$, $Cu[AlBr_4]$, and $Tl[AlBr_4]$. Further studies will be necessary in order to detect directly the existence of the ions $AsBr_2^+$ and $AsBr_4^-$ in solution. Unlike the other halides mentioned here, antimony(III) chloride is a fairly good solvent for alkali metal halides owing to its relatively high dielectric constant. For example, caesium, rubidium, potassium, ammonium, thallium, and quaternary ammonium chlorides, potassium bromide, and potassium fluoride are readily soluble, as are mercury(II) iodide, bromide, and chloride. The chlorides of lithium, sodium, tin(II), bismuth(III), and iron(III) dissolve only very slightly. Salts of oxo acids, with the exception of tetramethylammonium sulphate and perchlorate, are insoluble or react with the solvent.

The conductivity of antimony(III) chloride is again explained by the dissociation

$$2\,SbCl_3 \rightleftharpoons SbCl_2^+ + SbCl_4^-.$$

Chlorides such as aluminium(III) chloride and antimony(V) chloride gave solutions having good conductivities, which probably contain the ions $SbCl_2^+$ and $AlCl_4^-$ or $SbCl_6^-$. The chlorides may therefore be regarded as acids (chloride ion acceptors):

$$SbCl_3 + AlCl_3 \rightleftharpoons SbCl_2^+ + AlCl_4^-.$$

Tetramethylammonium chloride and even triphenylmethyl chloride also give conducting solutions, which probably contain the anion $SbCl_4^-$, so that these chlorides should be regarded as bases (chloride ion donors). Conductimetric titrations between acids and bases (neutralizations) have been carried out. Antimony(III) sulphate also reacts with the base tetramethylammonium chloride in a neutralization reaction. The conductivity curve contains salient points at molar ratios of 1:6 and 4:6 in agreement with the following equilibria:

$$6\,(CH_3)_4N[SbCl_4] + Sb_2(SO_4)_3 \rightleftharpoons 3\,[(CH_3)_4N]_2SO_4 + 8\,SbCl_3$$
$$3\,[(CH_3)_4N]_2SO_4 + 3\,Sb_2(SO_4)_3 \rightleftharpoons 6\,(CH_3)_4N[Sb(SO_4)_2].$$

Solvolyses have also been observed in antimony trichloride. For example, when silver perchlorate dissolves in this solvent, silver chloride precipitates out:

$$AgClO_4 + SbCl_3 \longrightarrow SbCl_2[ClO_4] + AgCl\downarrow.$$

Potassium bromide and potassium iodide give a product having the composition $2KCl \cdot SbCl_3$. Potassium fluoride on the other hand is simply solvated, and forms the compound $KF \cdot 2SbCl_3$. Oxides sulphides, carbonates, and acetates give the corresponding chlorides:

$$3\,M^{II}O + 2\,SbCl_3 \longrightarrow 3\,M^{II}Cl_2 + Sb_2O_3.$$

Organic substances are also dissolved and ionized by antimony(III) chloride. For example, organic chlorine compounds in concentrated solution dissociate in accordance with

$$2\,RCl \rightleftharpoons R_2Cl^+ + Cl^-,$$

whereas in dilute solution they dissociate as follows:

$$RCl \rightleftharpoons R^+ + Cl^-.$$

Aromatic hydrocarbons give strongly coloured solutions containing carbonium ions in antimony(III) chloride.

The ions $SbCl_2^+$ and $SbCl_4^-$ postulated on the basis of conductimetric measurements in antimony(III) chloride solutions have not yet been directly detected. Support for the occurrence of a process of the type

$$SbCl_4^- + {}^*SbCl_3 \rightleftharpoons SbCl_3 + {}^*SbCl_4^-$$

is provided by the high mobility of the Cl^- ion in this solvent [as in arsenic(III) chloride].

10.1.3. Mercury(II) Bromide

The melting points of arsenic(III) bromide and antimony(III) chloride are above room temperature. Mercury(II) bromide is a solvent having an even higher melting point. The discussion of this solvent will show that ionizing water-like solvents are also to be found among the melts of half salts. Salt melts, which no longer correspond to the definition of water-like solvents, since they are largely dissociated, will be discussed in this series by W. Sundermeyer.

Fused mercury bromide, which is a light yellow liquid between 238–320 °C is a very good solvent for inorganic and organic substances (Table 29). Some of the solutions have appreciable conductivities.

Table 29. Solubilities in fused mercury(II) bromide

Solubility	Element or compound
Abundantly soluble	$Hg(ClO_4)_2$, $HgBrClO_4$, $HgCl_2$, HgI_2, $Hg(CN)_2$, $Hg(SCN)_2$, HgO, HgS (black) Hg_2Br_2, $Hg(NO_3)_2$, $NaBr$, KBr, $RbBr$, $CsBr$, NH_4Br, $(CH_3)_4NBr$, $TlBr$, $AgBr$, $CuBr_2$, $CuBr$, $AlBr_3$, $SbBr_3$, $AgNO_3$, $TlNO_3$, $TlClO_4$, K_3PO_4, Tl_3PO_4, $K[PbBr_3]$, $K_2[PbBr_4]$, $K_2[HgI_4]$, $Cu_2[HgI_4]$, $K_2[Hg(CN)_4]$, S, Se, Te
Fairly abundantly soluble	$HgSO_4$, HgS (red) $HgSe$, $LiBr$, alkaline earth metal bromides, $ZnBr_2$, $CdBr_2$, $PbBr_2$, $CoBr_2$, NH_2HgBr, Hg_2SO_4, $(NH_4)_2SO_4$, $Pb(ClO_4)_2$, KIO_3
Insoluble	$KClO_4$, sulphates of Ag, Cd, K, Li, Ni, Pb, Sb(III), $Pb(NO_3)_2$, AlF_3, Al_2O_3, $NaPO_3$, Na_3PO_4, V_2O_5, TiO_2, ZrO_2, $Hg(IO_3)_2$, $Hg_3(PO_4)_2$.

Some of these salts form adducts with mercury (II) bromide, e.g. $KBr \cdot HgBr_2$, $2KBr \cdot HgBr_2$, $4KBr \cdot HgBr_2$, $8KBr \cdot HgBr_2$, $2\,HgO \cdot HgBr_2$, and $3\,HgO \cdot HgBr_2$, The low electric conductivity of fused mercury bromide is explained by bromidotropy:

$$2\,HgBr_2 \rightleftharpoons HgBr^+ + HgBr_3^-\ .$$

Substances that can provide $HgBr^+$ ions are thus acids (bromide ion acceptors), while those that can supply $HgBr_3^-$ ions are bases (bromide ion donors). Acids and bases of this type are assumed to exist on the grounds of conductivity measurements in the solutions of salts in fused mercury(II) bromide:

$$Hg(ClO_4)_2 + HgBr_2 \rightleftharpoons 2\,HgBr(ClO_4)$$

$$Hg(NO_3)_2 + HgBr_2 \rightleftharpoons 2\,HgBr(NO_3)$$

$$KBr + HgBr_2 \rightleftharpoons K[HgBr_3]\ .$$

Mercury(II) sulphate, which is only slightly soluble, is a weak acid. Examples of bases in mercury bromide are potassium bromide, sodium bromide, ammonium bromide, thallium bromide, lead bromide, and the bromides of other metals. Mercury oxide is an ansolvo base:

$$HgO + 2\,HgBr_2 \rightleftharpoons HgOHgBr^+ + HgBr_3^-.$$

Many neutralizations of the following type are known:

$$(HgBr)ClO_4 + K[HgBr_3] \longrightarrow KClO_4\uparrow + 2\,HgBr_2.$$

The arrow pointing upward indicates that owing to the high specific gravity of mercury(II) bromide, potassium perchlorate does not sediment, but collects at the surface of the melt. Neutralizations of this type can also be followed conductimetrically. In the titration of mercury perchlorate with thallium bromide, the conductivity passes through a minimum at a molar ratio of 1 : 2:

$$Hg(ClO_4)_2 + 2\,TlBr \longrightarrow 2\,TlClO_4 + HgBr_2$$

No direct (e.g. spectroscopic) evidence of the existence of the ions $HgBr^+$ and $HgBr^-$ in mercury bromide has yet been found, though these ions appear to present the best explanation for the experimental data.

10.1.4 Oxide Halides

In addition to the halides mentioned, there are a few oxide halides that may be regarded in many respects as water-like solvents. These include carbonyl chloride (phosgene) $COCl_2$, nitrosyl chloride $NOCl$, the important phosphoryl chloride (phosphorus oxide trichloride) $POCl_3$, and seleninyl chloride $SeOCl_2$. Phosgene is a poor solvent, and is difficult to handle because of its sensitivity to moisture, its chemical reactivity and its toxicity. Its low conductivity is attributed to self-ionization in accordance with

$$COCl_2 \rightleftharpoons COCl^+ + Cl^-$$

i.e. by assuming that phosgene is a chloridotropic solvent. Acids and bases are therefore all chloride ion acceptors and donors respectively, according to the ionotropic theory, or all substances that increase the concentrations of $COCl^+$ and Cl^- ions respectively according to the solvent theory. The existing experimental data are mainly concerned with the behaviour of aluminium chloride, which gives solutions having appreciable conductivities in phosgene. Aluminium chloride may be regarded as an ansolvo (chloride ion acceptor) acid:

$$AlCl_3 + COCl_2 \rightleftharpoons AlCl_3 \cdot COCl_2 \rightleftharpoons COCl^+ + AlCl_4^-.$$

These solutions attack metals such as potassium, magnesium, calcium, zinc, and cadmium, which are practically unaffected by the pure solvent:

$$Ca + COCl^+ \longrightarrow CO\uparrow + Ca^{2+} + Cl^-.$$

The compound $CaCl_2 \cdot 2AlCl_3 \cdot 2COCl_2$ can be isolated from solutions of calcium and $AlCl_3$ in phosgene. Since chlorides are bases in phosgene, the reaction of aluminium chloride with metal chlorides, e.g. calcium chloride, is a neutralization:

$$2\,COCl^+ + 2\,AlCl_4^- + Ca^{2+} + 2\,Cl^- \longrightarrow Ca(AlCl_4)_2 + 2\,COCl_2.$$

Nitrosyl chloride, NOCl, is also regarded as a chloridotropic solvent on the basis of its conductivity, which is explained by the equation:

$$NOCl \rightleftharpoons NO^+_{\text{solvated}} + Cl^-_{\text{solvated}}.$$

It has mainly been used as a solvent for nitrosyl compounds such as $NO[AlCl_4]$, $NO[FeCl_4]$, and $NO[SbCl_6]$. Solutions of these compounds are good conductors of electricity, and the compounds themselves must therefore be regarded as acids according to the solvent and ionotropic theories, though their structure in the solid state is not known for certain and they could be solvates of aluminium, antimony, and iron chlorides. The NO^+ cation has a remarkably high mobility, which is probably due to a chain conduction mechanism similar to that of the proton in water or of the chloride ion in arsenic(III) or antimony(III) chloride. The reactions of such nitrosyl compounds with chlorides (e.g. tetramethylammonium chloride) in nitrosyl chloride are neutralizations, and give the classical V curve in conductimetric titrations.

The most important oxide halide with solvent properties is the widely used and investigated phosphoryl chloride $POCl_3$. Its liquid range is readily accessible, and the compound is easy to purify and has good solvent properties for a series of inorganic compounds. For example, silicon(IV) chloride and bromide and tin(IV) bromide dissolve without dissociation or association, whereas tetraalkylammonium halides, phosphorus(V) chloride, phosphorus(V) bromide, and arsenic(III), bismuth(III), iodine(III), and tin(IV) chlorides dissociate on dissolution. Alkali metal and ammonium halides are only slightly soluble. The high value of 0.8 has been found for the transport number of the chloride ion in solutions of tetramethylammonium chloride; this is probably due to a chloride ion transfer from $POCl_4^-$ ions to $POCl_3$ molecules.

This chloride transfer is in agreement with the self-ionization

$$2\,POCl_3 \;\rightleftharpoons\; POCl_2^{+}{}_{(solv)} + POCl_4^{-}{}_{(solv)}.$$

proposed to explain the slight conductivity of the solvent.

However, the postulated chloridotropy has not been detected in some cases and can therefore be used only with reservations as a basis for an acid-base definition (cf. Section 7.4). Thus exchange experiments with radioactive chlorine in the system boron trichloride-phosphoryl chloride showed that rapid chlorine exchange occurs in the presence of excess phosphoryl chloride, but not with excess boron trichloride. The ions $POCl_2^{+}$ and BCl_4^{-}, which should be expected if boron trichloride behaves as an ansolvo acid, i.e. as a chloride ion acceptor, and whose presence appears to be confirmed by conductimetric experiments (see below), cannot exist in appreciable quantities in solution. In the solutions containing excess phosphoryl chloride, the exchange may proceed via the following steps:

$$POCl_3 \;\rightleftharpoons\; POCl_2^{+} + Cl^{-}$$

$$\overset{*}{Cl}{}^{-} + Cl_3B \cdot OPCl_3 \;\rightleftharpoons\; (Cl\overset{*}{Cl}_3B \cdot OPCl_3)^{-} \;\rightleftharpoons\; \overset{*}{Cl}{}^{-} + Cl_3B \cdot OPCl_3.$$

A solid adduct $BCl_3 \cdot POCl_3$, in which the boron trichloride is probably joined to the phosphoryl chloride via the oxygen, has in fact been isolated. It is difficult, however, to explain why the solvated tetrachloroborate(III) ion is not also formed in solutions containing excess boron trichloride. The conductivity of boron trichloride in phosphoryl chloride could be explained by the following dissociation:

$$Cl_3B \cdot OPCl_3 \;\rightleftharpoons\; Cl_3B \cdot OPCl_2^{+} + Cl^{-}.$$

However, this dissociation is not compatible with the curve for the conductimetric titration of boron trichloride with tetramethylammonium chloride in phosphoryl chloride. This curve contains a salient point, and indicates a dissociation in accordance with

$$BCl_3 + POCl_3 \;\rightleftharpoons\; POCl_2^{+} + BCl_4^{-}.$$

Further experiments will therefore be necessary in order to explain the equilibria that occur.

The two molecules in the solid adduct $SbCl_5 \cdot POCl_3$ are again linked by the oxygen atom. The results of conductimetric titration with tetramethylammonium chloride suggest the presence of the ions $POCl_2^{+}$ and $SbCl_6^{-}$ in solution:

$$POCl_2[SbCl_6] + (CH_3)_4\,NCl \longrightarrow POCl_3 + (CH_3)_4N[SbCl_6],$$

though dissociation equilibria involving cationic antimony have also been suggested:

$$Cl_3POSbCl_5 \rightleftharpoons Cl_3POSbCl_4^+ + Cl^- \rightleftharpoons Cl_2\overset{+}{P}OSbCl_5 + Cl^-.$$

The solid adducts of phosphoryl chloride with aluminium and gallium chlorides are probably also coordinated via the oxygen. The nature of solutions of aluminium chloride in phosphoryl chloride is uncertain. Titrations with tetramethylammonium chloride give a conductivity curve with a salient point at a molar ratio of 1:1 (formation of the compound $(CH_3)_4N[AlCl_4]$. The conductimetric titration of tin (IV) chloride with aluminium chloride gives a titration curve having a salient point at a molar ratio of 1:2 and an insoluble compound having the composition

$$SnCl_4 \cdot 2\,AlCl_3 \cdot 2\,POCl_3.$$

The system aluminium chloride/antimony(V) chloride/phosphoryl chloride is still more complex. Conductimetric titration of aluminium chloride with antimony(V) chloride gives a change in direction at a molar ratio of 1:2 ($AlCl_3 \cdot 2SbCl_5$); on titration of antimony(V) chloride with aluminium chloride, on the other hand, salient points are found at molar ratios of 1 : 1 ($SbCl_5 \cdot AlCl_3$), 1:2 ($SbCl_5 \cdot 2AlCl_3$), and possibly also 1:3 ($SbCl_5 \cdot 3AlCl_3$). These experiments suggest the presence of the anion $AlCl_4^-$, but also point to the simultaneous occurrence of cationic species such as $AlCl(OPCl_3)_5^{2+}$ and $Al(OPCl_3)_6^{3+}$.

Solid solvates having the composition $2FeCl_3 \cdot 3POCl_3$, $FeCl_3 \cdot POCl_3$, and $2FeCl_3 \cdot POCl_3$ can be isolated from the concentrated, red-brown solutions of iron (III) chloride in phosphoryl chloride, which contain no $FeCl_4^-$ ions. $FeCl_4^-$ ions are however present in the more dilute, yellow solutions. As is shown by the formation of the $FeCl_4^-$ ion from iron(III) chloride in triethyl phosphate, the chloridotropy of phosphoryl chloride in accordance with

$$FeCl_3 + POCl_3 \rightleftharpoons POCl_2^+ + FeCl_4^-$$

is not necessary for an understanding of the formation of $FeCl_4^-$ ions, since the disproportionation of solvated iron(III) chloride molecules would be sufficient (cf. Section 7.4):

$$2\,FeCl_3 + 4\,POCl_3 \rightleftharpoons FeCl_2(OPCl_3)_4^+ + FeCl_4^-.$$

The self-ionization of phosphoryl chloride is thus less important in the chemistry of this solvent than its Lewis base properties, which are due to the oxygen. The self-ionization of phosphoryl chloride also appears to be of no importance in the formation of $FeCl_4^-$ ions on addition of chlorides such as $(C_2H_5)_4NCl > KCl >$

$ZnCl_2 > PCl_5 > AlCl_3 > SbCl_5 > HgCl_2 > SnCl_4$. (The sign $>$ indicates a decreasing tendency of formation of the $FeCl_4^-$ ion.) The chloridotropy of the solvent is activated only on addition of strong Lewis bases such as pyridine or triethylamine:

$$C_5H_5N + POCl_3 + FeCl_3 \rightleftharpoons C_5H_5NPOCl_2^+ + FeCl_4^-.$$

This can also be seen from the fact that these Lewis bases react as ansolvo bases in the sense of the solvent theory, and give conducting solutions even when present alone:

$$(C_2H_5)_3N + POCl_3 \rightleftharpoons (C_2H_5)_3NPOCl_2^+ + Cl^-$$

Chemistry in seleninyl chloride $SeOCl_2$, as in phosgene, is largely determined by the high reactivity of this solvent, which is strongly oxidizing. Non-metals such as sulphur, phosphorus, selenium, and tellurium react readily with seleninyl chloride to form chlorides. Reaction is also observed with many metals. For example, the reaction with copper proceeds as follows:

$$3\ Cu + 6\ SeOCl^+ \longrightarrow 3\ Cu^{2+} + Se_2Cl_2 + 2\ SeO_2 + 2\ SeOCl_2$$

$$3\ Cu^{2+} + 6\ Cl^- \longrightarrow 3\ CuCl_2.$$

The chlorides of a number of metals such as lithium, potassium, rubidium, caesium, magnesium, calcium, strontium, and barium are inert and moderately soluble in the solvent. Tin (IV), antimony (V), and iron (III) chlorides are readily soluble, and form solvates having the compositions $SnCl_4 \cdot 2SeOCl_2$, $SbCl_5 \cdot 2SeOCl_2$, and $FeCl_3 \cdot 2SeOCl_2$.

The conductivity of pure seleninyl chloride is explained by the self-ionization

$$2\ SeOCl_2 \rightleftharpoons [SeOCl \cdot SeOCl_2]^+ + Cl^-$$

i.e. by the assumption that seleninyl chloride is a chloridotropic solvent. Though acids and bases in this solvent were originally defined according to the modified electronic theory (cf. Section 7.2), the reactions investigated and the equations given are not in conflict with the ionotropic and solvent theories. Tin(IV) chloride is an ansolvo acid in the sence of the solvent theory:

$$SnCl_4 + 2\ SeOCl_2 \rightleftharpoons 2\ SeOCl^+ + SnCl_6^{2-}.$$

Iron(III) chloride and sulphur trioxide behave similarly:

$$SO_3 + SeOCl_2 \rightleftharpoons SeOCl^+ + SO_3Cl^-$$

The latter is the strongest (ansolvo) acid in seleninyl chloride. Examples of ansolvo bases are ammonia, pyridine, and quinoline.

$$C_5H_5N + SeOCl_2 \rightleftharpoons C_5H_5NSeOCl^+ + Cl^-.$$

Isoquinoline is the strongest (ansolvo) base in seleninyl chloride. A neutralization that can be followed conductimetrically is represented by the following equation

$$2 C_5H_5NSeOCl^+ + 2 Cl^- + 2 SeOCl^+ + SnCl_6^{--}$$

$$\longrightarrow 2 C_5H_5NSeOCl^+ + SnCl_6^- + 2 SeOCl_2.$$

Neutralizations of the ansolvo bases pyridine, quinoline, and isoquinoline with the ansolvo acid sulphur trioxide proceed similarly; the curves correspond to those of other potentiometric titrations. Solutions of sulphur trioxide also dissolve oxides of aluminium, chromium, titanium, molybdenum, vanadium and uranium.

10.1.5. Dinitrogen Tetroxide

Liquid dinitrogen tetroxide N_2O_4 is a very poor solvent for inorganic ionic compounds because of its low dielectric constant. It is a better solvent for organic substances such as saturated and aromatic hydrocarbons, halogeno and nitro compounds, and carboxylic acids. Chemistry in this solvent is also determined by its Lewis acid character, its oxidizing action, and its ability to dissociate in several equilibria. For example, it exists both in the gas phase and in the liquid in equilibium with the brown, paramagnetic nitrogen dioxide:

$$N_2O_4 \rightleftharpoons 2 NO_2.$$

To explain the extremely low conductivity, the solvent is assumed to dissociate into nitrosyl and nitrate ions:

$$N_2O_4 \rightleftharpoons NO^+ + NO_3^-.$$

Finally, a number of reactions with organic compounds suggest that it also exists to a small degree in the equilibrium:

$$N_2O_4 \rightleftharpoons NO_2^+ + NO_2^-.$$

On the basis of the self-ionization of dinitrogen tetroxide into nitrosyl and nitrate ions, nitrosyl compounds such as nitrosyl chloride and nitrosyl bromide may be regarded as acids according to the solvent theory, and nitrates as bases. Metals such as iron, zinc, and tin react with "acidic" or even neutral solutions with liberation of nitric oxide:

$$M + 2NOCl \xrightarrow{\text{Liq. } N_2O_4} MCl_2 + 2NO \ (M = Zn, Fe, Sn \text{ etc}).$$

The solvates of some nitrates having the formulae $Zn(NO_3)_2 \cdot 2N_2O_4$, $Cu(NO_3)_2 \cdot N_2O_4$ and $Fe(NO_3)_3 . N_2O_4$ may be effectively "ansolvo acids", i.e. nitrosyl nitratometallates having the formulae $(NO)_2[Zn(NO_3)_4]$, $NO[Cu(NO_3)_3]$, and $NO[Fe(NO_3)_4]$. Strong Lewis bases such as tertiary amines form ionic adducts, which conduct electricity in solution and may be regarded as "ansolvo bases":

$$[(B)_nNO]^+[NO_3]^- \rightleftharpoons nB + N_2O_4 \rightleftharpoons N_2O_4 \cdot nB.$$

However, this is a formal view, since the conductivity of dinitrogen tetroxide is very low and, owing to the low dielectric constant of this solvent, even solutions of the "acid" nitrosyl chloride have only very low conductivites. Nitrosyl ions are therefore not present in appreciable quantities, and it is consequently more formal than accurate to consider these as the carriers of acid properties. Thus dinitrogen tetroxide like sulphur dioxide, is situated on the borderline between water-like and non-water-like solvents. The acid-base act can be described with the aid of the electronic theory, but since the solvent theory has proved suitable as a classifying principle for this solvent, it will be retained in the following discussion. An example of a "neutralization reaction" is represented by the following equation

$$NOCl + AgNO_3\downarrow \xrightarrow{\text{Liq. } N_2O_4} AgCl\downarrow + N_2O_4.$$

Many solvolyses and instances of amphoterism are known. For example, diethylammonium chloride is solvolysed:

$$(C_2H_5)_2NH_2Cl + N_2O_4 \rightleftharpoons NOCl + (C_2H_5)_2NH_2NO_3.$$

Reactions of this type are reversible. Their occurence depends on solubility of the chloride and the removal of the nitrosyl chloride from the system. Nitrosyl chloride can be obtained in preparative quantities by a similar reaction in which dinitrogen tetroxide is passed through a column containing moistened potassium chloride:

$$KCl + N_2O_4 \longrightarrow NOCl + KNO_3.$$

Magnesium perchlorate is solvolysed with formation of nitrosyl perchlorate:

$$Mg(ClO_4)_2 + 2N_2O_4 \longrightarrow Mg(NO_3)_2 + 2NOClO_4.$$

The solvolysis of aluminium chloride yields the chlorinefree adduct $Al(NO_3)_3 \cdot N_2O_4$, which should probably be formulated as $NO[Al(NO_3)_4]$. Lithium carbonate is also converted into the nitrate

$$Li_2CO_3 + 2 N_2O_4 \longrightarrow 2 LiNO_3 + N_2O_3 + CO_2$$

and other carbonates, hydroxides, and sulphides can be converted into anhydrous nitrates in the same manner.

An example of amphoteric behaviour is found in zinc nitrate, which is soluble in solutions of diethylammonium nitrate in dinitrogen tetroxide:

$$Zn(NO_3)_2 + x (C_2H_5)_2NH_2NO_3 \longrightarrow [(C_2H_5)_2NH_2]_x[Zn(NO_3)_{x+2}].$$

These peranamphoteric nitratozincates, the composition of which is not accurately known, can also be obtained by dissolution of zinc metal in a solution of diethylammonium nitrate in dinitrogen tetroxide:

$$Zn + x (C_2H_5)_2NH_2NO_3 + 2 N_2O_4 \longrightarrow [(C_2H_5)_2NH_2]_x[Zn(NO_3)_{x+2}] + 2 NO.$$

Many adducts of dinitrogen tetroxide with organic compounds, particularly ethers, are known. For example, 1,4-dioxan forms the adduct $O(CH_2CH_2)_2O \cdot N_2O_4$, which melts at 45.2 °C, and is colourless, crystalline, and diamagnetic. A different situation is found in the 1 : 1 and 2 : 1 adducts with the Lewis acid BF_3, which probably have the structures:

$$[NO_2^+] \begin{bmatrix} OBF_3^- \\ N \\ OBF_3 \end{bmatrix} \quad \text{and} \quad [NO_2^+] \begin{bmatrix} OBF_3^- \\ N \\ O \end{bmatrix}$$

The 2 : 1 adduct can be used as a reagent for the nitration of organic compounds.

10.2. Organic Solvents

In addition to acetic acid (cf. Section 5) and other anhydrous organic acids there are also available nowadays further organic solvents that may be regarded more or less as ionizing solvents. These include amines, nitriles, nitro compounds, amides, alcohols, ketones, acid anhydrides, sulphoxides, and sulphones. The physico-chemical properties of the most important of these compounds are listed in Table 30. They share the property of behaving more or less as electron donors (Lewis bases) toward the cationic component of the solute, i.e. the cations are solvated. This solvation then favours dissociation. The self-ionization of the solvents, on the other hand, is of practically no importance in the dissociation processes, so that the acid-base definition and the acid-base act cannot be defined in accordance with the solvent theory. The electronic theory of acids and bases is therefore used instead (cf. the coordination model, Section 7.4). Thus these solvents are again situated on the borderline with non-water-like solvents. The only exception are the unsubstituted amides and acetic anhydride, which may be regarded as water-like solvents because of their appreciable conductivity and self-ionization, and for which the solvent theory of acids and bases can therefore be used as a classifying principle.

Little is known about the use of primary amines such as methylamine as solvents. The primary amines are liquids whose low viscosity is considerably increased by the addition of soluble salts. They are poorer solvents for ionic compounds than ammonia. Salts such as lithium chloride, sodium nitrate, potassium thiocyanate, copper(I) thiocyanate, silver nitrate, silver iodide, strontium nitrate, barium thiocyanate, mercury cyanide, mercury iodide, and bismuth iodide are however readily soluble in methylamine, whereas heavy metal sulphates such as silver, copper(II), chromium(III) and nickel sulphates are insoluble. The soluble salts give conducting solutions. Tertiary amines do not appear to be capable of forming electrolytes. It should be mentioned that some primary amines like ammonia, can dissolve alkali metals to form blue solutions.

Of the diamines, only ethylenediamine has been investigated to any great extent as an ionizing solvent. It has a convenient liquid range, but also has the disadvantage that, like all amines, it readily absorbs moisture and carbon dioxide. The solubilities of various salts in ethylenediamine are lower than in liquid ammonia, but show the same trend. For example, the solubility of potassium iodide in ethylenediamine at 25 °C is 74.9 g per 100 g of solvent, whereas its solubility in liquid ammonia is 182 g per 100 g. Other readily soluble salts are sodium bromide, sodium iodide, sodium thiocyanate, sodium nitrate, sodium chlorate, sodium perchlorate, and potassium thiocyanate. Potassium perchlorate is slightly more soluble in ethylenediamine than in water. Chlorides are only very sparingly soluble. Many salts also form solvates. Examples are $NaI \cdot 3en$ and $SrCl_2 \cdot 6en$ (en = ethylenediamine). The numerous en complexes of transition metals need only be mentioned in passing. The soluble salts are dissociated in ethylendiamine, as is shown by the electrical conductivity of the solutions.

Ethylenediamine is used in analytical chemistry. For example, organic chlorine compounds (chloroform, dichloroethyl ether, etc.) react with the solvent at 100 °C to form secondary amines and amine hydrochlorides. The latter are "acids" in ethylenediamine, and can be titrated with "bases"

Table 30. Physico-chemical properties of some organic ionizing solvents

	Melting point (°C)	Boiling point (°C)	Dielectric constant	Specific conductivity (Ω^{-1} cm^{-1})	Viscosity (cPoise)	Density (g/cm^3)	Molecular weight
Methylamine	-93.5	-6.3	—	—	—	0.699 (-11 °C)	31.06
Ethylendiamine	11.0	116.2	12.9 (25 °C)	9×10^{-8}	1.725 (25 °C)	0.891 (25 °C)	60.1
Pyridine	-41.8	115.6	12.3 (25 °C)	$< 1 \times 10^{-9}$	0.829 (30 °C)	0.9728 (30 °C)	79.10
Acetonitrile	-45.7	81.6	36.2 (25 °C)	5.9×10^{-8} (25 °C)	0.325 (30 °C)	0.7768 (25 °C)	41.05
Nitromethane	-28.5	101.3	35.9 (30 °C)	6.56×10^{-7} (25 °C)	0.595 (30 °C)	1.1312 (25 °C)	61.04
Nitrobenzene	5.8	210.8	34.8 (30 °C)	9.1×10^{-7} (25 °C)	1.634 (30 °C)	1.193 (25 °C)	123.11
Formamide	2.6	193	111 (25 °C)	4×10^{-6} (25 °C)	3.31 (25 °C)	1.13 (25 °C)	45.04
N,N-Dimethylformamide	-61	153	36.7 (25 °C)	2×10^{-7} (25 °C)	0.796 (25 °C)	0.9443 (25 °C)	73.19
Acetamide	81	221	59.1 (83 °C)	4×10^{-6} (94 °C)	1.32 (105 °C)	0.990 (86,5 °C)	59.07
N-Methylacetamide	29.8	202	178.9 (30 °C)	1×10^{-7}–3×10^{-7} (40 °C)	3.019 (40 °C)	0.942 (40 °C)	73.09
N,N-Dimethylacetamide	-20	165	37.8 (25 °C)	0.8×10^{-7}–2×10^{-7} (25 °C)	0.919 (25 °C)	0.9366 (25 °C)	87.13
Urea	132.7	—	3.5 (22 °C)	—	—	1.335	60.06
Tetramethylurea	-1.2	176.5	23.1	0.6×10^{-7}	—	0.969 (20 °C)	116.16
Methanol	-97.5	64.5	32.6 (25 °C)	1.5×10^{-9} (25 °C)	0.5445 (25 °C)	0.7868 (25 °C)	32.04
Ethanol	-114.5	78.3	24.3 (25 °C)	1.35×10^{-9} (25 °C)	1.078 (25 °C)	0.7851 (25 °C)	46.07
Acetone	-95.4	56.2	20.7 (25 °C)	5.8×10^{-8} (25 °C)	0.2954 (30 °C)	0.7851 (25 °C)	58.08
Acetic anhydride	-73.1	139.5	22.1 (20 °C)	2×10^{-7}–5×10^{-7} (25 °C)	0.8511 (25 °C)	1.081 (20 °C)	102.09
Dimethylsulfoxide	18.4	189.0	46.6 (25 °C)	3×10^{-8} (20 °C)	1.9 (25 °C)	1.096 (25 °C)	78.13
Tetramethylensulfone	28.86	283	44 (30 °C)	2×10^{-8} (25 °C)	9.87 (30 °C)	1.2615 (30 °C)	120.17

Among the ammonia derivatives in the wider sense, pyridine has become particularly important as an ionizing solvent. In spite of its relatively low dielectric constant, it is a good solvent for inorganic substances, as can be seen from Table 31.

Table 31. Solubilities of inorganic substances in pyridine

Solubility	Compound
Abundantly soluble	$LiCl$, $LiBr$, LiI, $LiNO_3$, $NaBr$, NaI, $NaSCN$, $NaNO_2$, KCN, $KSCN$, $KMnO_4$, NH_4SCN, $CuCl$, $CuBr$, $CuSCN$, CuF_2, $AgSCN$, $AgNO_2$, $AgNO_3$, $AgCH_3COO$, Ag_2SO_4, $AuCl_3$, $BeCl_2$, $MgCl_2$, $MgBr_2$, MgI_2, $CaCl_2$, $CaBr_2$, CaI_2, $Ca(NO_3)_2$, $SrCl_2$, SrI_2, $BaCl_2$, BaI_2, $ZnCl_2$, $ZnBr_2$, $Zn(CN)_2$, CdI_2, $HgCl_2$, $HgBr_2$, HgI_2, $Hg(CN)_2$, $K_2[HgI_4]$, $AlBr_3$, $SnCl_2$, $SnCl_4$, $Pb(SCN)_2$, $PbCl_2$, $Pb(NO_3)_2$, $Pb(CH_3COO)_2$, TiF_4, $CrCl_2$, CrO_3, $MnCl_2$, $PtCl_4$, UBr_4, UO_2Cl_2, $FeCl_2$, $FeCl_3$, Br_2, I_2
Practically insoluble to insoluble	$LiOH$, $NaCl$, KOH, KNO_3, KCl, KBr, KI, NH_4Cl, NH_4I, $CuCN$, $Cu(NO_3)_2$, $BaBr_2$, $Ba(NO_3)_2$, $CdCl_2$, $CdBr_2$, $Hg(NO_3)_2$, $AlCl_3$, $PbBr_2$, PbI_2, $SbCl_3$, $BiCl_3$, $CrCl_3$, $Co(NO_3)_2$, $Ni(NO_3)_2$

Solutions of salts in pyridine are electrically conducting.
Pyridine is characterized by strong solvating power. The halogens bromine and iodine form solvates of the type $Br_2 \cdot Py$ and $I_2 \cdot Py$, in which charge-transfer bonding occurs. Iron (III) chloride dissolves presumably with formation of the solvate $FeCl_3 \cdot Py$ and cobalt(II) chloride forms the solvate $CoCl_2 \cdot 2\,Py$. The reaction of the solvate $FeCl_3 \cdot Py$ with the Lewis base Cl^- results in an equilibrium involving

the FeCl$_4^-$ anion. The salt CoCl$_2$ · 2 Py reacts similarly to give the anion [CoCl$_3$ · Py]$^-$. As a Lewis base, pyridine also dissociates the solvo acids of the aqueous system in accordance with

$$Py + HA \rightleftharpoons PyH^+ + A^-,$$

but owing to the low dielectric constant of the solvent, this dissociation is only slight: K = 7.5 x 10^{-4} for perchloric acid, 5.0 x 10^{-5} for nitric acid and 5.9 x 10^{-6} for hydrogen iodide. Nevertheless, pyridine is often used as a solvent for the determination of weak (organic) acids, since it intensifies the acidic character because of its Lewis basicity. Thus the readily enolizable β-diketones act as acids in pyridine, and can be titrated with sodium methoxide in the presence of thymolphthalein:

$$R-CO-CH_2-CO-R' + Py \rightleftharpoons R-CO-\overset{\ominus}{CH}-CO-R' + PyH^+.$$

Double decompositions and redox reactions are also possible in pyridine. For example, the reaction of silver nitrate with potassium thiocyanate gives a precipitate of potassium nitrate, and silver nitrate and barium iodide give a precipitate of barium nitrate; iron(III) chloride reacts with tin(II) chloride to give iron(II) chloride and tin(IV) chloride.

The best known organic ionizing solvent containing the nitrile group is acetonitrile. The solubilities of some inorganic compounds in acetonitrile are indicated in Table 32.

Table 32. Solubilities of inorganic substances in acetonitrile

Solubility	Compound
Abundantly soluble	LiBr, NH$_4$SCN, KSCN, CuBr$_2$, AgNO$_3$, CdI$_2$, HgCl$_2$, FeCl$_2$, FeCl$_3$, HCl, HBr, HI, H$_2$S, NH$_3$, SO$_2$, Cl$_2$, Br$_2$, I$_2$
Fairly abundantly to slightly soluble	LiCl, NaCl, NaBr, NaI, NaSCN, NH$_4$NO$_3$, KI, AlCl$_3$, SnCl$_2$, SnCl$_4$, CoCl$_2$, CoBr$_2$

It is possible to distinguish between strong (tetraalkylammonium halides, silver and potassium picrates, potassium iodide), medium-strong (silver nitrate, lithium and sodium picrates, monodi-, and trialkylammonium bromides and iodides), and weak electrolytes (mono-, di-, and trialkylammonium chlorides) in acetonitrile.

The predominant property of acetonitrile is again its solvating power. Solvates having various stoichiometries such as $CuBr_2 \cdot CH_3CN$, $ZnCl_2 \cdot 2\,CH_3CN$ and $CoCl_2 \cdot 3\,CH_3CN$ are known. The dissolution of cobalt(II) chloride in acetonitrile has been studied in detail, and can be described by the following equations:

$$2\,CoCl_2 + 6\,CH_3CN \longrightarrow Co(CH_3CN)_6{}^{2+} + CoCl_4{}^{2-}$$

$$CoCl_4{}^{2-} + 2\,CH_3CN \rightleftharpoons [CoCl_4(CH_3CN)_2]^{2-}.$$

Iron(III) chloride is also dissociated by solvation into cationic and anionic species

$$2\,FeCl_3 + 4\,CH_3CN \rightleftharpoons [FeCl_2(CH_3CN)_4]^+ + FeCl_4{}^-.$$

The same is true of phosphorus(V), arsenic(V), and antimony(V) fluorides and antimony(V) chloride

$$2\,PF_5 + 2\,CH_3CN \rightleftharpoons 2\,PF_5 \cdot CH_3CN \rightleftharpoons [PF_4(CH_3CN)_2]^+ + PF_6{}^-.$$

Phosphorus(V) chloride, on the other hand, dissociates without solvation

$$2\,PCl_5 \rightleftharpoons PCl_4{}^+ + PCl_6{}^-.$$

Solvation, which is so frequent because of the Lewis base properties of acetonitrile, also occurs with boron trifluoride which forms the compound $BF_3 \cdot CH_3CN$. Acetonitrile also acts as a Lewis base toward substances that release protons. Hydrogen chloride gives the unstable adducts $[CH_3CN \cdot H]\,Cl$ and $[CH_3CN \cdot H]\,HCl_2$. Perchloric acid, picric acid, dichloroacetic acid, mandelic acid, and acetic acid dissolve in acetonitrile to give solutions whose degrees of dissociation decrease in the order given. Perchloric acid can therefore be determined quantitatively in acetonitrile (with diphenylguanidine).

Acetonitrile also acts as a Lewis base in the following complex reactions. When Cl^- or $N_3{}^-$ ions are added to the perchlorates, tetrafluoroborates, tetrafluoroantimonates, and hexachloroantimonates of the ions Ti^{3+}, V^{3+}, VO^{2+}, Cr^{3+}, Mn^{2+}, and Fe^{3+} in acetonitrile, it is possible to detect not only the neutral chlorides or azides, but also the tetrachloro complexes $TiCl_4{}^-$ $VCl_4{}^-$, and $CrCl_4{}^-$ and the azido complexes $[Tl(N_3)_4]^-$, $[Ti(N_3)_6]^{3-}$, $[V(N_3)_6]^{3-}$, $[VO(N_3)_5]^{3-}$, $[Cr(N_3)_6]^{3-}$, $[Mn(N_3)_4]^{2-}$, and $[Fe(N_3)_6]^{3-}$. However, there are also cases in which the solvation is due to a Lewis acid action of acetonitrile. Thus it reacts with the Lewis base picoline to form the complex Picoline $\cdot CH_3CN$.

Formamide and acetamide, unlike the other organic solvents mentioned here, have appreciable conductivities, which must be interpreted as resulting from a self-ionization in accordance with

$$2\ RCONH_2\ \rightleftharpoons\ RCONH_3^+\ +\ RCONH^-$$

The solvent theory can therefore be used for the acid-base definition and the acid-base act in these two ionizing solvents. Both amides have a dielectric constant and are largely associated, as is shown by their relatively high viscosities. They are unusually good solvents for organic and inorganic compounds, and in this respect they strongly resemble water. Potassium perchlorate, the halides of mercury and lead, mercury oxide, and mercury phosphate are even more soluble in acetamide than in water. The solutions of inorganic compounds are electrically conducting. Electrolyses have also been carried out in these solvents. Their disadvantages are their sensitivity to moisture and other reactivities, particularly in the case of formamide.
Of the two ionizing solvents N-methylacetamide and N,N-dimethylacetamide, which have convenient liquid ranges and are relatively inert, the former has been particularly extensively studied. It has a very high dielectric constant, and, like formamide and acetamide, is an unusually good solvent for organic and inorganic compounds. It is almost comparable in its universality with water. Electrochemical studies, particularly conductivity measurements, show that the dissolved electrolytes are largely dissociated.
Owing to their Lewis basicity, both solvents have strong solvating actions, but differ in this respect, as can be seen from the following complex reactions. Dissolution of iron(III) chloride in N,N-dimethylacetamide (DMA) leads to the following equilibria:

$$2\ FeCl_3\ +\ DMA\ \rightleftharpoons\ Fe_2Cl_6\ \cdot\ DMA$$

$$Fe_2Cl_6\ \cdot\ DMA\ +\ 3\ DMA\ \rightleftharpoons\ [FeCl_2(DMA)_4]^+\ +\ FeCl_4^-.$$

The system $CoCl_2$=DMA contains the ions $CoCl_4^{2-}$, $[Co(DMA)_6]^{2+}$, $[CoCl_3(DMA)]^-$, and $[CoCl(DMA)_5]^+$, and in low concentrations, the species $CoCl_2(DMA)_2$ or $CoCl_2(DMA)_4$. N-methylacetamide (NMA), on the other hand, is characterized by stronger solvation of the anions, probably by means of hydrogen bonds of the type $X^-\ldots.H\ldots.N(CH_3COCH_3)$. Solutions of iron(III) chloride in NMA therefore contain only the complex cation $[Fe(NMA)_6]^{3+}$ and solvated Cl^-.
Acetone, methanol, and ethanol have been studied mainly in mixtures with water, and little experimental data is available on chemistry in the pure solvents. They are however strongly solvating solvents, and their self-ionization is of no importance. Acetone and the alcohols differ in their solvating properties in the same manner as N,N-dimethylacetamide and N-methylacetamide. Whereas acetone mainly solvates the cations, and so allows e.g. the formation of halogenocobaltates on dissolution

of cobalt(II) halides, the alcohols can also solvate halide-anions because of their tendency to form hydrogen bonds, with the result that e.g. the dissolution of iron(III) chloride in alcohols leads to the formation of solvated chloroiron cations and solvated chloride ions.

Acetic anhydride is fairly widely used as an ionizing solvent. It has a convenient liquid range, and is a good solvent for inorganic and in particular for organic compounds (Table 33).

Table 33. Solubilities of inorganic substances in acetic anhydride

Solubility	Compound
Abundantly soluble	$K(CH_3COO)$, $Rb(CH_3COO)$, $Ca(CH_3COO)_2$, $Si(CH_3COO)_4$, $Ge(CH_3COO)_4$, CH_3COCl, CH_3COBr, NaI, $NaClO_4$, $AgClO_4$, $MgBr_2$, $SbBr_3$, $SbCl_5$
Slightly soluble	$Na(CH_3COO)$, $[N(CH_3)_4](CH_3COO)$, $Ba(CH_3COO)_2$, $Zn(CH_3COO)_2$, $LiClO_4$, KI, $[N(CH_3)_4]Cl$, $MgCl_2$, $ZnCl_2$, $HgBr_2$, $SnCl_2$
Insoluble	$Cu(CH_3COO)_2$, $Ag(CH_3COO)$, $Mg(CH_3COO)_2$, $Hg(CH_3COO)_2$, $Ni(CH_3COO)_2$, $AgCl$, $TlCl$, $TlBr$, TlI, Ag_2S, PbS, As_2S_3

Solvates of some acetates are known, a few examples being $Na(CH_3COO) \cdot (CH_3CO)_2$ $K(CH_3COO) \cdot (CH_3CO)_2O$, $2 Rb(CH_3COO) \cdot 1(CH_3CO)_2O$, $MgBr_2 \cdot 6 (CH_3CO)_2O$, $2 NaI \cdot CdI_2 \cdot 6(CH_3CO)_2O$.

The conductivity of very pure acetic anhydride is explained by the self-ionization

$$(CH_3CO)_2O \rightleftharpoons CH_3CO^+ + CH_3COO^-.$$

This dissociation scheme is supported by exchange experiments with labelled (^{11}C, ^{14}C) compounds. Labelled sodium acetate, labelled acetic acid, and labelled acetyl chloride undergo isotope exchange with acetic anhydride, in some cases fairly rapidly. On the basis of this dissociation scheme, according to the solvent theory, substances that increase the acetyl ion concentration in acetic anhydride are acids, and substances that increase the acetate ion concentration are bases. Acetyl compounds act as solvo acids, and acetates as solvo bases. They enter into many neutralization reactions as follows:

$$CH_3CO^+ + CH_3COO^- \longrightarrow (CH_3CO)_2O.$$

These reactions can be very readily followed conductimetrically, since solutions of solvo bases and salts (but not solvo acids) in acetic anhydride have appreciable conductivities. Some examples are given in the following equations:

$$CH_3COCl + K(CH_3COO) \longrightarrow KCl\downarrow + (CH_3CO)_2O$$

$$CH_3COBr + Tl(CH_3COO) \longrightarrow TlBr\downarrow + (CH_3CO)_2O$$

$$CH_3COSCN + Rb(CH_3COO) \longrightarrow RbSCN + (CH_3CO)_2O$$

$$(CH_3CO)_2S + (CH_3COO)_2Pb \longrightarrow PbS\downarrow + 2\,(CH_3CO)_2O$$

$$CH_3CO(C_6H_5SO_3) + CH_3COO[N(CH_3)_4] \longrightarrow$$
$$[(CH_3)_4N]SO_3C_6H_5 + (CH_3CO)_2O.$$

When acetyl chloride is added to potassium acetate in acetic anhydride, potassium chloride precipitates out. The conductivity decreases and reaches a constant value at a molar ratio of 1 : 1. If the salts formed remain in solution, the conductivity curves have a different shape, but are still easy to evaluate. Neutralizations in acetic anhydride can also be followed potentiometrically. This possibility has recently been used for the determination of very weak nitrogen bases (ansolvo bases).

A considerable volume of data is also available on solvolyses in acetic anhydride. Metal carbonates liberate carbon dioxide even in the cold

$$Na_2CO_3 + (CH_3CO)_2O \longrightarrow 2\,Na(CH_3COO) + CO_2\uparrow.$$

Sulphites behave similarly:

$$K_2SO_3 + (CH_3CO)_2O \longrightarrow 2\,K(CH_3COO) + SO_2\uparrow.$$

Many sulphides are also solvolysed. Thus sodium sulphide gives acetyl sulphide:

$$Na_2S + 2\,(CH_3CO)_2O \longrightarrow 2\,CH_3COONa + (CH_3CO)_2S.$$

The solvolyses of nitrates and nitrites are very complex. In addition to the acetates they yield oxides of nitrogen, nitrogen itself, and carbon dioxide, which result from the decomposition of the intermediate acetyl nitrate or nitrite. The solvolyses of halides of non-metals have also been studied. Silicon tetrachloride gives silicon(IV) acetate:

$$SiCl_4 + 4\,(CH_3CO)_2O \rightleftharpoons Si(CH_3COO)_4 + 4\,CH_3COCl.$$

This solvolysis can be followed conductimetrically by addition of thallium acetate, which binds the acetyl chloride:

$$4\,Tl(CH_3COO) + 4\,CH_3COCl \longrightarrow 4\,(CH_3CO)_2O + 4\,TlCl\!\downarrow.$$

After the addition of 4 moles of thallium acetate per mole of silicon tetrachloride, the conductivity shows no further change. Phosphorus(V) chloride, phosphorus oxide trichloride, and phosphorus(III) chloride are also solvolysed in acetic anhydride, but the solvolysis products have not yet been identified. The first stage in the reaction of phosphorus(V) chloride is

$$PCl_5 + (CH_3CO)_2O \longrightarrow POCl_3 + 2\,CH_3COCl.$$

However, phosphorus oxide chloride undergoes further solvolysis. Phosphorus(III) chloride reacts as follows:

$$PCl_3 + 3\,(CH_3CO)_2O \rightleftharpoons 3\,CH_3COCl + P(CH_3COO)_3.$$

This brief discussion of organic ionizing solvents will be concluded with a reference to dimethyl sulphoxide (DMSO) and tetramethylene sulphone (TMSO$_2$). Dimethyl sulphoxide dissolves simple salts such as alkali metal halides, tetraalkylammonium halides and picrates, and alkali metal and ammonium nitrates, thiocyanates, and perchlorates. These are all largely dissociated and give conducting solutions. Iron(III) chloride forms cationic complexes of the type Fe (DMSO)$_6^{3+}$. The tetrachloroferrate ion is unstable in DMSO. Thus DMSO has stronger solvating properties (is a stronger Lewis base) than e.g. N,N-dimethylacetamide, in which the tetrachloroferrate ion is stable. An interesting situation is found in solutions of nickel(II) chloride in DMSO. Green solutions of hexacoordinate complexes, probably Ni(DMSO)$_6^{2+}$ or NiCl(DMSO)$_5^+$, are stable at room temperature. When heated to 50 °C, the solutions turn blue, presumably as a result of the formation of tetrahedral complexes. The colour change is reversible. On prolonged heating at 50 °C, blue crystals having the composition [Ni (DMSO)$_6$]NiCl$_4$ separate out. This compound is also obtained at room temperature on addition of benzene. The complexes that can be isolated as solids and those that exist in solution at the same temperature are thus sometimes quite different.

Tetramethylene sulphone resembles liquid sulphur dioxide in its good solvent properties for inorganic compounds. The spectrum of a solution of iron (III) chloride in $TMSO_2$ shows that the tetrachloroferrate ion is formed. The solvating properties of $TMSO_2$ are thus weaker than those of DMSO. Cobalt (II) chloride and cobalt (II) perchlorate dissolve to give blue and red solutions respectively, but the conductivities of the solutions are low, indicating that the compounds are not appreciably dissociated. A blue solid having the composition $CoCl_2 \cdot TMSO_2$ can be isolated from the blue solutions of cobalt(II) chloride; on the basis of optical studies, this complex is assumed to have a tetrahedral structure. A red solid having the composition $Co(TMSO_2)_3$ $(ClO_4)_2$ isolated from the red solutions of cobalt(II) perchlorate appears from optical studies to resemble the complex $Co(H_2O)_6{}^{2+}$ in its structure.

10.3. Bibliography

Arsenic(III) fluoride, chloride, bromide, Antimony(III) chloride:
    H. Spandau u. V. Gutmann: Angew. Chem. **64**, 93 (1952)
    D.S. Payne: Halides and Oxyhalides of Group V Elements as Solvents, in T.C. Waddington: Non-Aqueous Solvent Systems. Academic Press, London, New York 1965, Chapter 8
Mercury(II) bromide:
    G. Jander u. K. Brodersen: Z. anorg. allg. Chem. **261**, 261 (1950); **262**, 33 (1950); **264**, 57, 76, 92 (1951); **265**, 117 (1951)
    H. Spandau u. V. Gutmann: Angew. Chem. **64**, 93 (1952)
Carbonyl-, Nitrosyl-, Phosphoryl-, Seleninyl chloride:
    H. Spandau u. V. Gutmann: Angew. Chem. **64**, 93 (1952)
    L.F. Audrieth u. J. Kleinberg: Non-Aqueous Solvents. Wiley, New York 1953, Chapter 12
    D.S. Payne: Halides and Oxyhalides of Group V Elements as Solvents, in T.C. Waddington: Non-Aqueous Solvent Systems. Academic Press, London, New York 1965, Chapter 8
Dinitrogentetroxide:
    H. Spandau u. V. Gutmann: Angew. Chem. *64*, 93 (1952)
    C.C. Addison: ibid. **72**, 193 (1960)
    H.H. Sisler: Chemistry in Non-Aqueous Solvents, Reinhold, New York 1961
Aliphatic Amines, Pyridine, Acetonitrile:
    L.F. Audrieth u. J. Kleinberg: Non-Aqueous Solvents, Wiley, New York 1953, Chapter 7

G. Hohlstein u. U. Wannagat: Z. anorg. allg. Chem. **288,** 193 (1956)

R.S. Drago u. K.F. Purcell: Co-ordinating Solvents, in T.C. Waddington: Non-Aqueous Solvent Systems. Academic Press, London, New York, 1965, Chapter 5

V. Gutmann: Angew. Chem. **78,** 151 (1966)

L. Kolditz: ibid. **78,** 452 (1966)

Formamide, Acetamide, N-Methylacetamide, N,N-Dimethylacetamide

L.F. Audrieth u. J. Kleinberg: Non-Aqueous Solvents. Wiley, New York 1953, Chapter 7

G. Winkler: Chemie in geschmolzenem Acetamid, in G.Jander, H. Spandau u. C.C. Addison: Chemie in nichtwässerigen ionisierenden Lösungsmitteln, Vieweg, Braunschweig 1963, Vol. IV, part 3

L.R. Dawson: Chemistry in Formamide and Derivatives of Amides, in G. Jander, H. Spandau u. C.C. Addison: Chemie in nichtwässerigen ionisierenden Lösungsmitteln. Vieweg, Braunschweig 1963, Vol. IV, part 5

R.S. Drago u. K.F. Purcell: Co-ordinating Solvents, in T.C. Waddington: Non-Aqueous Solvent Systems. Academic Press, London, New York 1965, Chapter 5

Urea, Tetramethylurea:

B. Sansoni: Angew. Chem. **66,** 596 (1954)

A. Lüttringhaus u. W. Dirksen: ibid. **75,** 1059 (1963)

Methanol, Ethanol, Acetone:

R.S. Drago u. K.F. Purcell: Co-ordinating Solvents, in T.C. Waddington: Non-Aqueous Solvent Systems. Academic Press, London, New York 1965,Chapter 5

Acetic anhydride:

G. Jander: Die Chemie in wasserähnlichen Lösungsmitteln. Springer, Berlin, Göttingen, Heidelberg 1949, Chapter 9

H. Surawski: Chemie in Essigsäureanhydrid, in G. Jander, H. Spandau a. C.C. Addison: Chemie in nichtwässerigen ionisierenden Lösungsmitteln. Vieweg, Braunschweig 1963, Vol. IV, Part 2

Dimethylsulfoxide, Tetramethylensulfone:

H.L. Schläfer u. W. Schaffernicht: Angew. Chem. **72,** 618 (1960)

R.S. Drago u. K.F. Purcell: Co-ordinating Solvents, in T.C. Waddington: Non-Aqueous Solvent Systems. Academic Press, London, New York 1965, Chapter 5

11. Subject Index

ChT 3

J. Jander-
Ch. Lafrenz
Ionizing Solvents

Dr. Jochen Jander (born 1925 in Göttingen) is professor of ino
ganic chemistry at the Free University of Berlin. After studyir
chemistry at the universities of Greifswald and Göttingen he
was awarded his doctorate for a thesis on the red color of ru
based on work carried out in Professor Thilo's laboratory.
He then spent several years with Professor Emeléus in
Cambridge and in 1957 received his venia legendi for a thesis
on compounds of nitrogen with monovalent electronegative
components. These investigations have since been continued
with particular regard to preparative and structural aspects.
After five years at the Technische Hochschule in Munich,
Professor Jander took up his present post at Berlin in 1967.
Dr. rer. nat. Christian Lafrenz was born 1937 in Burg/Fehm
(Germany). After studying chemistry at the universities of Ki
Innsbruck (Austria), Freiburg, and Munich he received his
doctorate for a thesis on bromine-nitrogen compounds. He th
spent one year as postdoctoral research fellow with Professc
R. W. Parry at the University of Michigan. Today he works in 1
research institute of the BP Company.

ISBN 0 471 43970 3